ENVIRONMENTAL SCIENCE
The Way the World Works

STUDY GUIDE

Clark E. Adams
Bernard J. Nebel

THIRD EDITION

ENVIRONMENTAL SCIENCE

The Way the World Works

Bernard J. Nebel
Catonsville Community College

PRENTICE HALL, *Englewood Cliffs, New Jersey 07632*

Editorial/production supervision: **Maureen Lopez**
Interior design: **Clark Adams**
Manufacturing buyer: **Paula Massenaro**

A Division of Simon & Schuster
Englewood Cliffs, New Jersey 07632

Printed in the United States of America

10 9 8 7 6 5 4 3

ISBN 0-13-282229-6

Prentice-Hall International (UK) Limited, *London*
Prentice-Hall of Australia Pty. Limited, *Sydney*
Prentice-Hall Canada Inc., *Toronto*
Prentice-Hall Hispanoamericana, S.A., *Mexico*
Prentice-Hall of India Private Limited, *New Delhi*
Prentice-Hall of Japan, Inc., *Tokyo*
Simon & Schuster Asia Pte. Ltd., *Singapore*
Editora Prentice-Hall do Brasil, Ltda., *Rio de Janeiro*

TABLE OF CONTENTS

CHAPTER 11
TOXIC CHEMICALS AND GROUNDWATER POLLUTION

CHAPTER 12
AIR POLLUTION AND ITS CONTROL

CHAPTER 13
ACID PRECIPITATION, THE CO_2 GREENHOUSE EFFECT, AND DEPLETION OF THE OZONE SHIELD

CHAPTER 19
ENERGY RESOURCES AND THE NATURE OF THE ENERGY PROBLEM

CHAPTER 20
NUCLEAR POWER, COAL, AND SYNTHETIC FUELS

CHAPTER 21
SOLAR ENERGY, OTHER "RENEWABLE ENERGY SOURCES", AND CONSERVATION

CHAPTER 22
LIFESTYLE, LAND USE AND ENVIRONMENTAL IMPACT

TO THE STUDENT

HELPFUL HINTS ON HOW TO USE THIS STUDY GUIDE

The authors of this study guide attempted to develop it by playing the role of the student who would be using it. For example, you will note that the study guide topics are correlated with and in the same sequence as they are presented in the text. So as you read each chapter of the text, you should easily find the answers to each question. However, this will not be the case if you are using an earlier edition of Environmental Science: The Way the World Works. This study guide was developed specifically for the 3rd edition of this text. The study guide can be used with earlier editions of this text but the transition from your reading to answering questions will be less natural than using the 3rd edition.

Each chapter of the study guide was constructed with three questions in mind, i.e., where are the students heading, how will they get there, and how will they know they have arrived? The introductory statement tells you where you are going by highlighting the particular issues and problems addressed in each chapter. The study questions were constructed to give you a directed system of learning thus minimizing wasted effort. Your success on the self tests should give you some indication of whether you have indeed arrived at one level of mastery of chapter content.

There are two ways you can use this study guide that will help you reinforce text content. One way is to first read an entire chapter and then attempt to answer as many of the study guide questions as possible. The correlations of study guide and textbook chapters and subsections should make it easy to look up the answers to questions. Any answers you need to look up require another reading of this section of the text and will alert you to content areas that need your additional attention. After you have attempted to answer all study questions, check your answers with the key at the end of the chapter. Finally, answer the self test questions and grade yourself by checking your answers against the self test answer key at the end of the chapter.

Another way to use this study guide is to answer each study question as you read the text. First read the study guide question, then read the text until you find the answer to that particular question. You will find that using this system will, at times, require that you read several pages before the answer to a particular question is revealed. This system of learning works because it focuses your study on one issue at a time. After completing all study guide questions, check your responses with those of the key at the end of the chapter. Complete your study of each chapter by answering the self test questions.

Remember that incorrect responses to either the study guide or self test questions are as valuable indicators of your mastery of chapter content as are correct answers. The incorrect responses alert you to specific areas that need additional attention and thus focuses your study time.

After you have all the correct answers to all study guide and self test questions, what else needs to be done? The questions represent the body of information that needs to be integrated into the context of humankind achieving ecosystem stability. Concentrated study of the information in the text and study guide will give you a comprehensive understanding of the pertinent ecological principles, how various human activities are in conflict with these principles, and how this conflict might be mitigated by redirecting human activities to be in concert with ecological principles. If you can address each issue on these three levels, then you should enjoy a high level of personal satisfaction and academic success.

ENVIRONMENTAL SCIENCE
The Way the World Works

INTRODUCTION

WHAT IS SCIENCE?
WHY STUDY ENVIRONMENTAL SCIENCE?

This chapter describes the scientific method, the method which we use to gain a realistic, factual understanding of the things and events we observe. This understanding includes an accurate knowledge of what things are, how they work and what causes them. Armed with such understanding, we are in a much better position to accomplish goals and solve problems that may arise. Could we have sent people to the Moon without such understanding? Can we hope to find a cure for AIDS without such understanding?

We all live in the Earth's environment. Our goal is to have an environment that is free from pollution, has ample resources for everyone, and supports abundant wildlife. In short we want an environment that is safe and healthy for ourselves, our children and for other plants and animals. However, we are aware that there are many environmental problems including water and air pollution, depletion of water resources, and loss of wildlife among others. These environmental problems are created by human society and must be solved by us. As members of human society using resources and producing wastes, we are, to some degree, part of the problem, but we can also be part of the solutions. It is a tremendous opportunity!

In order to solve these problems we need to apply the scientific method, first as a means of understanding how the environment functions, then as a means of understanding the problems and how to solve them. We should look at learning about the scientific method as learning how to use a tool which is indispensable for effective problem solving.

STUDY QUESTIONS

INTRODUCTION

1. Our current relationship with the environment (Is, Is Not) sustainable.

2. Human activities such as obtaining resources and manufacturing products need to be measured in terms of ________________ impacts in terms of what are acceptable ________________.

3. The 1960s and 70s are recognized as the era of the ________________ **movement.**

4. The 1980s was a time of ________________ and growing ________________.

5. The 1990s may be an era that takes us from nonsustainable development to a course of ________________ development.

6. The concept of sustainable development is to meet present needs without compromising those of the ________________.

7. Studies on how plants and animals interact with each other and their environment is called (Ecology, Environmental Science).

8. Studies of ecological principles and their application to the human condition is called (Ecology, Environmental Science).

9. Is it (True, False) that people's *values* are different and sometimes inconsistent?

10. Any movement toward sustainable development will require public

    a. understanding and awareness of the problems. (True, False)
    b. adoption of new values. (True, False)
    c. demands for change. (True, False)

EVALUATING INFORMATION

WHAT IS SCIENCE?

11. Scientific information is based on ________________ and ________________ deductions from observations.

12. The art of compiling scientific information is a series of steps called the ________________ **method.**

The Scientific Method

13. The scientific method begins with

    a. observations, b. developing hypotheses, c. experimentation, d. formulation of theories.

Observations and Facts

14. All scientific information is based on ________________.

15. In science, the only observations accepted are those made with the basic five senses, namely ______________, ________________, ________________, ________________, and ________________.

16. Before it is accepted, an observation must be confirmed or ________________.

17. Observations which cannot be confirmed are (Accepted, Rejected) as scientific fact.

18. Observations that are verified and confirmed are (Accepted, Rejected) as scientific fact.

19. Indicate whether the following statements are [+] or are not [-] correct.

    [ ] All observations are facts.
    [ ] All facts are verified observations.

20. Which of the following is the most fundamental feature of science, i.e., no information could be gained without it?

    a. making observations, b. performing experiments, c. using instruments

Hypotheses and Their Testing

After making an observation, one is often drawn to ask further questions about it such as: What caused it? Where did it come from? How does it work?

21. To obtain answers to such questions the scientist first makes an educated guess called a ________________.

22. The next step is to ________________ the hypothesis to see whether it is consistent or inconsistent with other observations.

23. The purpose of a scientific experiment is to:

    a. test a hypothesis, b. prove that a hypothesis is correct, c. confuse the public
    d. use fancy instruments, e. have fun at public expense

24. A hypothesis that cannot be tested is:

    a. correct, b. incorrect, c. useful, d. useless

25. A hypothesis is:

    a. an experiment, b. a tentative guess, c. a theory

Controlled Experiments

26. To give definitive results, an experiment must involve a minimum of ________ groups.

27. The two groups are called the ________________ group and the ________________ group.

28. The ________________ group is subjected to treatment or condition which the scientist believes is responsible for the observed effect.

29. The control group

    a. must be an exact replicate in every respect,
    b. must be the same except for the factor being tested,
    c. is any other group

30. Is it (True, False) that in the absence of a control group, one can be sure than an observed effect is due to the test treatment?

31. A group of plants is grown on a medium without potassium (a chemical treatment). The plants do poorly. The result of this experiment:

    a. proves that potassium is required for plant growth.
    b. proves that potassium is not required for plant growth.
    c. proves nothing because there is no control.
    d. proves nothing because other factors could be responsible for the poor growth.
    e. Both c and d are correct.

32. The above experiment could be interpreted as showing that plants require potassium if and only if:

    a. it were repeated.
    b. a similar group of plants, grown under exactly the same conditions but with the addition of potassium grew well.
    c. conditions were more rigorously controlled.

33. The purpose of the control group is to

a. show that conditions can be controlled.
b. show that factors other than the test factors are not responsible for the observed effect.

Formulation of Theories

34. A concept that provides a logical explanation for a certain body of facts is called a ________________

35. Theories are

a. tested in much the same way as are hypotheses. (True, False)
b. subjected to "if...then..." reasoning. (True, False)
c. used to predict outcomes. (True, False)
d. subject to change. (True, False)

Principles and Natural Laws

36. When we observe that certain events always occur in a precisely predictable way (e.g., an unsupported object always falls) we define such behavior as a ________________ or **natural** ________________.

37. Principles and natural laws are:

a. the same as hypotheses.
b. the same as theories.
c. descriptions of the way certain things are always observed to behave.
d. laws passed by a council of scientists.

THE ROLE OF INSTRUMENTS IN SCIENCE

38. List three uses of instruments in conducting the process of science.

a. ____

b. ____

c. ____

39. Scientists do **NOT** use instruments to:

a. increase their power of observation.
b. control conditions in various experiments.
c. quantify observations.
d. avoid making observations.

SCIENCE AND TECHNOLOGY

40. Indicate which of the below statements describe the goal of [a] **basic** or **pure science** and which describes [b] technology.

[ ] The application of scientific knowledge to gain a specific objective.
[ ] The search for knowledge and understanding for its own sake.

41. Is it (True, False) that an understanding of basic science benefits technology by reducing the trial-and-error approach?

42. Technology benefits basic science by

a. increasing scientists' powers of observation. (True, False)
b. serving as a proving ground for new theories and principles. (True, False)
c. reducing the need for observations. (True, False)

43. A new invention is proposed which violates basic natural laws and scientific theory. Such an "invention" is most likely to be a (Success, Failure).

UNRESOLVED QUESTIONS AND CONTROVERSY

44. Indicate whether the following statements are [+] or are not [-] accurate concerning the process of science.

[ ] There are no controversies or arguments among scientists.
[ ] Progress in science can be slow.
[ ] No hypotheses or theories are incorrect.
[ ] The process of science may lead to answers but also more questions.
[ ] Science is incapable of providing absolute proof for any theory.
[ ] The process of science can be used to test value judgments.
[ ] The validity of science is based on the ability to do experiments.

45. The reason for controversy in science is most likely to be:

a. dispute over facts, b. because alternative hypotheses have not been adequately tested.

SCIENCE AND VALUE JUDGMENTS

46. Indicate whether the following statements about the process of science are true [+] or false [-].

[ ] Science can provide understanding that will enable us to achieve various objectives.
[ ] Science can tell us which objectives to achieve.
[ ] Ultimately, science can give us answers to all questions.
[ ] Science can predict the outcome of various courses of action.
[ ] Science can predict those courses of action that ignore established principles and theories and that will end in failure.

ENVIRONMENTAL SCIENCE AND ITS APPLICATION

47. The study of how plants and animals interact with each other and their environment is known as ____________________.

48. Is it (True, False) that ecological principles and theories apply only to wild plants and animals?

49. The application of ecological principles and theories to the human situation is known as ____________________ science.

50. If the human system, involving agriculture and industry, among other enterprises, is to be sustainable over the long term, is it (True, False) that it must be operated according to ecological theory and principles?

51. When we examine the human system, is it (True, False) that we find that everything is in accordance with ecological principles?

52. Is it (True, False) that the root of all our environmental problems may be the failure to appreciate and abide by ecological principles and theories?

53. Bringing the human system into accordance with ecological principles and thereby assuring sustainability will depend on understanding the value judgments of:

    a. scientists only, b. politicians only, c. politicians and scientists, d. society as a whole

---

KEY VOCABULARY AND CONCEPTS

control group
environmental science
experimental group
experiments
fact
hypothesis
natural law
observations
principles
role of instruments
science
scientific method
scientific controversy
technology
test
theory
value judgment
verification

SELF TEST

Circle the correct answer to each question.

1. Studies of how plants and animals interact with each other and their environment is called

    a. cytology, b. ecology, c. zoology, d. environmental science

2. Studies of ecological principles and their application to the human condition is called

    a. cytology, b. ecology, c. zoology, d. environmental science

3. Any movement toward sustainable development will require public

    a. understanding and awareness of the problems.
    b. adoption of new values.
    c. demands for change.
    d. all of the above

4. The scientific method begins with

    a. observations, b. developing hypotheses, c. experimentation, d. formulation of theories

5. All scientific information is based initially on

    a. observations, b. developing hypotheses, c. experimentation, d. formulation of theories

6. To obtain answers to questions, scientists first

a. make observations, b. develop hypotheses, c. perform experiments, d. develop theories

7. Hypotheses are confirmed through

a. observations, b. experiments, c. authority figures, d. technology

8. Experiments must include

a. experimental groups, b. control groups, c. controlled environments, d. all of these

9. The purpose of the control group is to

a. show that conditions can be controlled.
b. show that factors other than the test factors are not responsible for the observed effect.
c. prevent contamination of the experimental group.
d. verify the beliefs of authority figures.

10. Theories are

a. tested in much the same way as hypotheses.
b. subjected to "if...then..." reasoning.
c. subject to change.
d. all of the above.

11. Events that always occur in a precisely predictable way are called

a. observations, b. hypotheses, c. theories, d. principles

12. Scientists do not use instruments to

a. increase their power of observation.
b. control conditions in various experiments.
c. quantify observations.
d. avoid making observations.

13. The application of scientific knowledge to gain a specific objective is the goal of

a. technology, b. basic science

14. Which of the following is not an accurate statement concerning the process of science?

a. The process of science may lead to answers and more questions.
b. The validity of science is based on the ability to do experiments.
c. There are no controversies or arguments among scientists.
d. Progress in science can be slow.

15. Which of the following is not an accurate statement concerning science?

a. Science can test value judgments.
b. Science can test one-time events.
c. Science can give us answers to all questions.
d. All of the above are not accurate.

ANSWERS TO STUDY QUESTIONS

1. Is Not; 2. environmental, tradeoffs; 3. environmental; 4. consolidation, professionalism; 5. sustainable; 6. future; 7. ecology; 8. environmental science; 9. true; 10. all true; 11. observations; logical; 12. scientific; 13. a; 14. observations; 15. sight, smell, taste, touch, hearing; 16. verified; 17. rejected; 18. accepted; 19. -, +; 20. b; 21. hypothesis; 22. test; 23. a; 24. d; 25. b; 26. two; 27. experimental, control; 28. experimental; 29. b; 30. false; 31. e; 32. c; 33. b; 34. theory; 35. all true; 36. principle, law; 37. c; 38. increase power of observation, control experimental conditions, quantify observations; 39. d; 40. b, a; 41. true; 42. true, true, false; 43. failure; 44. -, +, -, +, +, -, +; 45. b; 46. +, -, -, +, +; 47. ecology; 48. false; 49. environmental; 50. true; 51. false; 52. true; 53. d.

ANSWERS TO SELF TEST

1. b; 2. d; 3. d; 4. a; 5. a; 6. b; 7. b; 8. d; 9. b; 10. d; 11. d; 12. d; 13. a; 14. c; 15. d

CHAPTER 1

ECOSYSTEMS: WHAT ARE THEY?

A grouping of plants and animals interacting with each other and their environment in such a way that the whole grouping continues generation after generation is defined as an *ecosystem*. Humans, along with their agricultural crops and animals, industry, transportation, and urbanized lifestyles, represent a *human ecosystem*. In turn, air pollution, water pollution, depletion of resources, loss of wildlife, and other environmental problems can be taken as various malfunctions of the human ecosystem which we wish to correct.

If you want to repair a car it is necessary to first understand how cars work. Similarly, if we wish to repair the human ecosystem and make it function better, it is necessary to understand how ecosystems work. In Chapter 1 we will study various *natural ecosystems* to discover the basic parts and how they fit together to make a working system. We will note how the environment tends to shape the kind of ecosystem that exists in a region but the fundamental parts and the way they work is the same. Thus the terminology and concepts presented will give a basic foundation for discussing and understanding all ecosystems including the human ecosystem.

STUDY QUESTIONS

DEFINITION AND EXAMPLES OF ECOSYSTEMS

1. An ecosystem may be defined as a grouping of __________________ and __________________ interacting with __________________ and their __________________.

2. Furthermore, the interrelationships are such that the grouping may be __________________.

3. A grouping of plants and animals that interact with each other and their environment in such a way that the grouping is perpetuated is an __________________.

4. List six examples of major kinds of ecosystems.

   a. ____________________, b. ____________________, c. ____________________

   d. ____________________, e. ____________________, f. ____________________

5. Each ecosystem has a distinctive plant __________________.

6. A plant community consists of (One, Several) species of plants growing together in an area.

7. A species refers to a population of organisms that can do what? __________________.

8. Each distinctive plant community supports a more or less distinctive community of __________________ and also __________________.

9. The study of ecosystems and the interactions which occur in them is the science of __________________.

10. Very large ecosystems that generally contain variations and smaller ecosystems within them are referred to as ________________.

11. The major biome of eastern United States is ________________.

12. The major biome of central United States is ________________.

13. The major biome of southwestern United States is ________________.

14. Is it (True, False) that individual ecosystems and biomes have sharp boundaries?

15. Is it (True, False) that what occurs in one ecosystem or biome may affect other ecosystems or biomes?

16. All the ecosystems and biomes of the earth are interconnected into one overall system which is called the ________________.

STRUCTURE OF ECOSYSTEMS

17. The total array of plants, animals, and microbes in an ecosystem is referred to as the ________________.

18. The biota of a region refer to the total array of ________________, ________________, and ________________.

BIOTIC STRUCTURE

19. All the organisms in an ecosystem can be assigned to one of three categories. These three categories are ________________, ________________, and ________________.

Categories of Organisms

20. Producers include all plants that do what? ________________________________

21. In photosynthesis, plants use ________________ energy to convert ________________ gas and ________________ into ________________.

22. The word inorganic refers to substances such as ________________.

23. The chemicals that make up the bodies of organisms are referred to as ________________.

24. In short, producers (photosynthetic plants) produce (Inorganic, Organic) chemicals.

25. As raw material for this production, producers use (Inorganic, Organic) chemicals.

26. The energy for this production comes from ________________.

27. Organisms which can produce all their own organic materials from inorganic raw materials in the environment are called (Autotrophs, Heterotrophs).

28. Organisms which must consume organic material as a source of nutrients and energy are called (Autotrophs, Heterotrophs).

29. Is it (True, False) that **all** plants are autotrophs? If you said false, give an example of an exception. ________________

30. All organisms that must consume organic matter for survival are called:

a. producers, b. consumers, c. decomposers

31. List three organisms that could be classified as consumers.

a. ________________, b. ____________________, c. __________________

32. Consumers that feed directly on producers are called primary consumers or ________________.

33. Consumers that feed on other consumers are called secondary consumers or ________________.

34. An animal which is commonly killed and eaten by another is called the (Predator, Prey) in a ________________/prey relationship.

35. An organism which attaches to another and feeds on it over a period of time, generally harming but not killing it immediately, is called a ______________ in a (what kind of?) ________________ relationship.

36. All dead organic matter (dead leaves, twigs, bodies of animals, fecal matter) is called _________________.

37. Animals that feed primarily on detritus are called ___________________.

38. Give three examples of detritus feeders.

a. __________________, b. __________________, c. ___________________

39. Much of the detritus in an ecosystem appears to rot or decompose as opposed to being eaten by detritus feeders. Rotting, decay, or decomposition is really the result of the feeding action of organisms called:

a. producers, b. consumers, c. decomposers

40. Decomposers are members of what two groups of organisms? ________________ and ____________________.

Feeding Relationships: Food Chains, Food Webs, and Trophic Levels

41. A sequence of specific organisms, each of which may feed on the next in turn, is referred to as a food (Chain, Web).

42. Organisms may be grouped into feeding levels, namely producers, primary consumers (herbivores), secondary consumers (carnivores), and so on. Such groupings are referred to as a food (Chain, Web).

43. Producers are the (First, Second, Third) trophic level.

44. The total combined weight of any group of organisms is known as its _________________.

45. In an ecosystem the biomass of herbivores is always (Greater, Less) than the biomass of producers, and the biomass of carnivores is always (Greater, Less) than the biomass of herbivores.

46. The biomass relationship at each trophic level can be depicted as a:

a. box, b. rectangle, c. circle, d. pyramid

47. Feeding relationships between two different species in which both benefit from the relationship is called ________________.

48. Describe an example of mutualism.

____

____

49. Feeding on different things, in different locations, and at different times minimizes ________________ between animal species.

ABIOTIC FACTORS

50. The chemical/physical (nonliving) factors of the environment are known as ______________ factors.

51. Three examples of abiotic factors are:

a. ________________, b. ________________, c. ________________

Optimum Zones Of Stress, Limits of Tolerance

52. For any factor that affects growth, the best amount or level to give is known as the ______________.

53. As the amount or level of any factor increases or decreases from the optimum, the plant or animal is increasingly ________________.

54. If any factor increases or decreases beyond an organism's limit of tolerance, what occurs? ________________.

55. Is it (True, False) that each species has an optimum zone of stress, and limits of tolerance with respect to every factor?

Law of Limiting Factors

56. How may factors need to be beyond a plant's limit of tolerance to preclude its growth? ________

57. If a single factor is responsible for preventing the growth of a plant, that factor could be referred to as the ________________ **factor**.

58. Is it (True, False) that if only one factor is outside the optimal range, it will cause stress and limit the organism?

WHY DO DIFFERENT REGIONS SUPPORT DIFFERENT ECOSYSTEMS?

CLIMATE

59. The two climatic factors responsible for delimiting the major terrestrial biomes are ________________ and ________________.

60. ________________ is the main climatic factor responsible for the separation of ecosystems into forests, grasslands, and deserts.

61. Even though it takes 30 inches or more of rainfall to support a forest ecosystem, ______________ will determine the kind of forest.

62. Ground that remains permanently frozen is called ______________.

OTHER ABIOTIC FACTORS AND MICROCLIMATE

63. A localized area within an ecosystem that has moisture and temperature conditions different from the overall climate of the area is called a ______________.

BIOTIC FACTORS

64. The biotic factor that limits grass from taking over high rainfall regions is

a. tall trees, b. herbivores, c. parasites, d. insects

PHYSICAL BARRIERS

65. The spread of species into other ecosystems may be prevented by physical barriers such as ________________, ________________, and ________________.

BIOTIC AND ABIOTIC INTERACTIONS

66. Is it (True, False) that all ecosystems are maintained by a delicate interplay of limiting factors affecting all the species?

67. Is it (True, False) that altering any factor, abiotic or biotic, will invariably affect limits and set into motion a chain reaction with far reaching consequences?

IMPLICATIONS FOR HUMANS

68. As hunter-gatherers, humans were much like:

a. producers, b. consumers, c. decomposers

69. With ________________, humans created their own distinctive ecosystem apart from natural ecosystems.

70. List five ways humans have overcome the usual limiting factors that prevent the spread of our species.

a. ____

b. ____

c. ____

d. ____

e. ____

71. With the proliferation and spread of the human ecosystem, is it (True, False) that we are upsetting and destroying other ecosystems?

KEY WORDS AND CONCEPTS

perpetuate	ecosystem	deciduous forest
conifer forest	savanna	grassland
desert	plant community	tropical rain forest
species	population	microbes
biota	ecology	biome
biosphere	climate	abiotic factors
optimum	optimal	limiting factors
range of tolerance	limits of tolerance	biotic factors
producers	consumers	decomposers
photosynthesis	carbon dioxide	minerals
organic	inorganic	autotrophs
-troph-	heterotroph	primary consumer
herbivore	secondary consumer	carnivore
predator	prey	parasite
detritus	detritus feeder	food chain
food web	trophic levels	biomass

SELF TEST

Circle the correct answer to each question.

1. An ecosystem is best defined as

   a. a grouping of plants and animals that interact with each other and their environment in such a way that the grouping is perpetuated.
   b. any grouping of plants and animals.
   c. plants, animals, and environmental factors.
   d. environmental factors affecting plants and animals.

2. Ecosystems are generally characterized and named by their

   a. climatic conditions.
   b. dominant plant type.
   c. dominant animals.
   d. land forms.

3. Very large ecosystems that generally contain variations and smaller ecosystems within them are called

   a. biota, b. microclimates, c. biomes, d. habitats

4. A major biome in the central United States is a

   a. deciduous forest, b. grassland, c. desert, d. coniferous forest

5. The total array of plants, animals, and microbes in an ecosystem is referred to as the

a. ecosystem, b. biome, c. community, d. biota

6. Which of the following is not correctly matched?

a. oak tree - producer
b. squirrel - consumer
c. mushroom - detritus
d. fungi - decomposer

7. The three major biotic components of ecosystem structure are

a. producers, herbivores, and carnivores
b. producers, consumers, and decomposers
c. plants, animals, and climate
d. consumers, detritus feeders, and decomposers

8. Grass eaten by grasshoppers, that are eaten by frogs, that are eaten by snakes, that are eaten by owls is a description of a

a. food chain, b. food wed, c. trophic levels, d. biomass

9. The first trophic level refers to

a. sunlight, b. green plants, c. herbivores, d. carnivores

10. The combined weight of any group of organisms is known as

a. food chain, b. food web, c. trophic levels, d. biomass

11. The biomass relationship at each trophic level can be depicted as a

a. box, b. rectangle, c. circle, d. pyramid

12. Which of the following is not an example of an abiotic factor?

a. water, b. light, c. dead log, d. temperature

13. The points at which a factor becomes so high or low as to threaten an organism's survival is referred to as

a. range of optimums, b. range of limits, c. range of tolerance, d. limits of tolerance

14. How many factors need to be beyond an organism's limits of tolerance to cause stress?

a. one, b. five, c. ten, d. hundreds

15. The limiting factor that is responsible for the separation of ecosystems into forests, grasslands, or deserts is

a. temperature, b. water, c. pH, d. soil type

ANSWERS TO STUDY QUESTIONS

1. plants, animals, each other, environment; 2. perpetuated; 3. ecosystem; 4. deciduous forest, grassland, deserts, coniferous forest, tundra, tropical rain forest; 5. community; 6. several; 7. reproduce; 8. animals, microbes; 9. ecology; 10. biome; 11. deciduous forest; 12. grassland; 13. desert; 14. false; 15. true; 16. biosphere; 17. biota; 18. plants, animals, microbes; 19. producers, consumers, decomposers; 20. carry on photosynthesis; 21. light, carbon dioxide, water, glucose; 22. air, rocks, or water; 23. organic; 24. organic; 25. inorganic; 26. light; 27. autotrophs; 28. heterotrophs; 29. false, chemotrophs; 30. b; 31. herbivores, carnivores, omnivores; 32. herbivores; 33. carnivores; 34. prey, predator; 35. parasite, symbiotic; 36. detritus; 37. detritus feeders; 38. vultures, earthworms, termites; 39. c; 40. fungi, bacteria; 41. chain; 42. web; 43. first; 44. biomass; 45. less, less; 46. d; 47. mutualism; 48. lichens, flowers and insects; 49. competition; 50. abiotic; 51. light, temperature, water; 52. optimum; 53. stressed; 54. death; 55. true; 56. one; 57. limiting; 58. true; 59. temperature, water; 60. water; 61. temperature; 62. permafrost; 63. microclimate; 64. a; 65. ocean, desert, mountain range; 66. true; 67. true; 68. b; 69. agriculture; 70. producing abundant food, creating reservoirs and distributing water, overcoming predation and disease, constructing our own habitats, overcoming competition with other species; 71. true

---

ANSWERS TO SELF TEST

1. a; 2. b; 3. c; 4. b; 5. d; 6. c; 7. b; 8. a; 9. b; 10. d; 11. d; 12. c; 13. d; 14. a; 15. b

CHAPTER 2

ECOSYSTEMS: HOW THEY WORK

In Chapter 1, we pointed out the observations that no species on Earth lives in isolation as an entity unto itself. Instead, every species is found in the context of a grouping of various plants and animals interacting with each other and their environment. We have defined such a grouping as an ecosystem. We have seen that the ecosystem is effectively shaped by the environment such that different ecosystems occupy different climatic regions. Despite apparent differences, every ecosystem contains plants, animals, and microbes that fit into definable categories, namely producers, consumers, and decomposers. Further, we have noted that the kinds of feeding and other relationships among organisms are essentially the same regardless of the kind of ecosystem being studied.

This chapter explains ecosystem structure and function. Ecosystem structure is explained in terms of identifying those atoms (N, CHOPS) most used to build the living (biotic) component of ecosystems. Ecosystem function is discussed in terms of how producers and consumers use energy and how nutrients are recycled. Chapter content may appear difficult because there is no personal recognition of things like atoms, molecules, matter, and energy in your day-to-day life. Additionally, students have a hard time grasping how energy is brought into the ecosystem, how it is used, stored, and lost. Ecosystems work because matter (N, CHOPS) is recycled through processes driven by a never-ending supply of solar energy. However, ecosystems use matter and energy differently. Energy use is a one-way trip! It comes into the ecosystem as light energy, captured through photosynthesis, and stored in green plants as glucose. Both plants and animals rely on this initial supply of glucose for all life support functions and activities. Both use cellular respiration to release the energy stored in glucose. All energy is ultimately lost from the ecosystem in the form of heat. If you can grasp the principles of recycling matter, energy flow, and biomass, the concepts presented in this and succeeding chapters will make sense. In fact, you will (should) be shocked by the implications of human ignorance regarding these principles.

STUDY QUESTIONS

In order to gain an understanding of how ecosystems function to propagate and perpetuate themselves it is necessary to have some knowledge of basic chemical concepts. Therefore, if you are not familiar with the ideas of atoms, elements, molecules, chemical bonding, and chemical reactions, you may want to first go through the set of study questions pertaining to Appendix C.

APPENDIX C: SOME BASIC CHEMICAL CONCEPTS

ATOMS, ELEMENTS, AND COMPOUNDS

1. The word "matter", in a chemical sense, refers to all ________________, ________________, and ________________.

2. All gases, liquids, and solids (as opposed to things like light and heat) are referred to as ________________.

3. All matter is comprised of fundamental chemical particles called __________________.

4. If the atoms comprising a substance are of one kind, the substance is called an __________________.

5. The word element is used to refer to material which is comprised of one specific kind of __________________.

6. If a substance is comprised of two or more different kinds of atoms, it is called a __________________.

7. The word "compound" refers to substances which are comprised of __________________ kinds of atoms.

8. How many different kinds of atoms exist in nature? __________________

9. Indicate whether each of the following is an element [E] or compound [C].

   [ ] oxygen
   [ ] carbon dioxide
   [ ] water
   [ ] nitrogen
   [ ] air
   [ ] phosphorus
   [ ] carbon
   [ ] protein
   [ ] sugar
   [ ] hydrogen

ATOMS, BONDS, AND CHEMICAL REACTIONS

10. Indicate whether the following statements about chemical reactions are true [+] or false [-].

   [ ] Atoms may be destroyed.
   [ ] New atoms may be created.
   [ ] An atom of one element may be changed into an atom of another element.
   [ ] Atoms are rearranged in the way they are bonded together.
   [ ] Atoms are never created, destroyed, or changed from one into another.

   The facts pointed out in the above question are so absolute that they are often referred to as the **Law of Conservation of Matter.**

The Structure of Atoms

11. All atoms have the same basic structure of __________________, __________________, located in the nucleus and __________________ around the outside.

12. However, atoms differ in the number of __________________ found in the nucleus of the atom.

13. In a neutral atom, the number of electrons equals the number of __________________.

14. One atom has 6 protons: another has 8. These two atoms (May be the same, Must be different) elements. The two elements are (see figure C-1) __________________ and __________________.

15. Two atoms with the same number of protons but different numbers of neutrons (Will be the same, May be different) elements. Such atoms are referred to as different ________________ of the same element.

Bonding of Atoms

16. The two basic ways in which atoms are bonded together are through ________________ and ________________ bonding.

17. The electrons around an atom are arranged in a specific layer or ________________ that has a (Specific, Random) number of spaces for electrons.

18. The first orbital closest to the nucleus can hold (One, Two, Eight); the next can hold (One, Two, Eight).

19. Is it (True, False) that electrons are arranged such that an inner orbital must be filled before any electrons go into the next orbital?

20. An atom is most stable when its outermost orbital is (Completely, Incompletely) filled or completely ________________.

21. The number of positive and negative charges must be the (Same, Different).

22. Two atoms may satisfy these two conditions, maintaining a balanced charge and completion of orbitals, by doing what? ____

Covalent Bonding

23. A bond formed by sharing one or more pairs of electrons is called a (Ionic, Covalent) bond.

24. A grouping of two or more atoms held together by covalent bonds is called a ________________.

25. Atoms of what element can form long chains and rings through covalent bonding? ________________

26. Atoms of what other elements also form covalent bonds readily? ________________, ________________, ________________, ________________, and ________________.

27. All the complex molecules which make up the tissues of living things are based on ________________ bonding between the above atoms.

Ionic Bonding

28. Atoms can also achieve a stable arrangement of electrons by ________________ one or more electrons to complete the filling of their outermost orbital or by ________________ one or more electrons to empty the orbital.

29. An atom which has gained or lost one or more electrons is called an ________________.

30. An atom which has gained one or more electrons is a (Positive, Negative) ion.

31. An atom which has lost one or more electrons is a (Positive, Negative) ion.

32. Positive and negative ions (Attract, Repel) each other.

33. Ionic bonding refers to atoms held together by ____________________.

34. An ionicly bonded substance must involve two or more different elements. Why?____________________.

35. Two examples of substances which are ionicly bonded are __________ and __________.

36. Ionicly bonded substances can be referred to as:

a. compounds, b. molecules, c. either

Chemical Reactions and Energy

37. Different bonding arrangements between different atoms have the (Same, Different) degrees of stability.

38. Chemical reactions involve the changing of bonding arrangements so that the atoms involved achieve (Greater, Less) stability.

39. As bonding arrangements change toward greater stability, __________ is released from the reaction.

40. Chemical reactions always tend to go in a direction that (Releases, Stores) energy.

41. Hydrogen gas reacts (burns explosively) with oxygen to form water. Why doesn't water spontaneously break up to form hydrogen and oxygen?

____

42. Is it (True, False) that chemical reactions can be made to go in a direction that yields less stable, higher energy products? If you said "true", give an example __________

43. What must be added to the system to make it go toward less stable, higher energy product?

____

ELEMENTS, LIFE, ORGANIZATION, AND ENERGY

ORGANIZATION OF ELEMENTS IN LIVING AND NONLIVING SYSTEMS

1. Different ecosystems have (Different, Basically the same) structural components and function in basically (Different, The same) way.

2. Living things contain (The Same, Different) elements that are found in the nonliving environment.

3. Six common elements in living things are __________, __________, __________, __________, __________, and __________.

4. Abiotic (nonliving) molecules consist of (Relatively Simple, Extremely complex) arrangements of atoms whereas molecular arrangements in living things are (Simple, Extremely complex).

5. Air is a simple mixture of (1, 2, 3, 4, 5) important gases. Name and give the chemical formula of each. ________________, ________________, and ________________

6. Water molecules contain two atoms of ________________ bonded to one atom of ________________.

7. For each of the following elements, which are critical in living organisms, give a compound or molecule which contains it and where it is found in the abiotic environment. (see Table 2-1)

Element	Compound Containing It	Environmental Location
carbon	________________	________________
hydrogen	________________	________________
oxygen	________________	________________
nitrogen	________________	________________
phosphorus	________________	________________

8. Indicate whether the following statements are true [+] or false [-] concerning the **Law of Conservation of Matter.**

[ ] atoms cannot be created
[ ] atoms can be destroyed
[ ] atoms can be converted into others
[ ] atoms can be rearranged to form different molecules

CONSIDERATIONS OF ENERGY

Matter and Energy

9. Matter is defined as anything that occupies ________________ and has ________________.

10. From the most to least dense states, matter can exist in the forms of ________________, ________________ and ________________.

11. Energy is defined as: ____

12. List three forms of energy. ________________, ________________, and ________________

13. Indicate whether the following statements about energy are true [+] or false [-].

[ ] Energy occupies space.
[ ] Energy can be weighed.
[ ] Kinetic energy can be transformed into potential energy.
[ ] Potential energy can be transformed into kinetic energy.
[ ] Energy can be recycled.
[ ] Energy can be created.
[ ] Energy can be converted from one form to another.
[ ] There is some loss in each energy conversion.
[ ] Energy can be measured.

14. Heat, light, and motion are forms of (Kinetic, Potential) energy.

15. Glucose, a compressed spring, and fossil fuels are forms of (Kinetic, Potential) energy.

16. The high potential energy in fuels is commonly referred to as ________________ energy.

17. A _________________ is unit measure of the amount of heat required to raise the temperature of 1 gram of water 1 degree Celsius.

18. Indicate whether the following statements are true [+] or false [-] concerning the **First Law of Thermodynamics.**

    [ ] Energy cannot be created or destroyed.
    [ ] Energy can be converted from one form to another.

19. Indicate whether the following statements are true [+] or false [-] concerning the **Second Law of Thermodynamics.**

    [ ] Energy conversions always result in net losses of useful energy.
    [ ] Heat is the final energy conversion form.

Energy and Organic Matter

20. Identify whether the level of potential energy in the following types of molecules is high [H] or low [L].

    [ ] inorganic [ ] organic [ ] coal
    [ ] water [ ] sugar [ ] wood

MATTER AND ENERGY CHANGES IN ORGANISMS

PRODUCERS

21. In most ecosystems, the primary producers are plants that carry on ________________.

22. Light energy for photosynthesis is absorbed by the __________________ molecule.

23. The efficiency of photosynthesis in converting light energy to chemical energy is at best ___________ percent efficient.

24. The primary product of photosynthesis is __________________.

25. The two basic functions of glucose are to provide raw __________________ for **plant** growth and __________________ for all **plant** functions.

CONSUMERS

26. The two basic functions of glucose are to provide raw __________________ for **animal** growth and __________________ for all **animal** functions.

27. The process through which organic molecules are broken down to release their potential energy is called **cellular** ____________________.

28. Cell respiration is effectively the _________________ process of photosynthesis in terms of what is consumed and released.

29. Indicate whether each of the following refers to photosynthesis [P], cell respiration [R], both [B], or neither [N].

[ ] releases oxygen
[ ] stores energy
[ ] releases carbon dioxide
[ ] consumes carbon dioxide
[ ] releases energy
[ ] produces sugar
[ ] consumes sugar
[ ] consumes oxygen

30. Place an [X] by any of the following organisms that carry on cell respiration for at least part of their energy needs.

[ ] consumers
[ ] decomposers
[ ] plants
[ ] fungi
[ ] bacteria

31. The lack of certain nutrients in your diet may lead to various diseases caused by ________________.

32. Eating excessive amounts of junk food at the expense of other more nutritive foods may also lead to various diseases caused by ________________.

33. The point of a balanced diet is that it supplies both ______________ and ________________ in adequate but not excessive amounts.

34. Food that is not digested passes through the digestive system and out as ________________ wastes.

Detritus Feeders and Decomposers

35. Dead leaves, stems, branches, and other woody organic material is called ________________.

36. Detritus consists mostly of ________________ that cannot be digested by most consumers.

37. Which of the following organisms are considered detritus feeders.

a. carnivores, b. fungi, c. bacteria, d. green plants, e. humans

PRINCIPLES OF ECOSYSTEM FUNCTION

NUTRIENT CYCLING

38. *Resources are supplied and wastes are disposed of by recycling all elements* is the (First, Second, Third) basic principle of ecosystem function.

39. Natural ecosystems do not harm themselves with their own waste products because these products are ________________.

40. Natural ecosystems do not run out of essential resources because these resources are ________________.

The Carbon Cycle

41. The reservoir of carbon for the carbon cycle is ________________ present in the ________________.

42. Carbon dioxide is first incorporated into sugar by the process of ________________ and then it may be passed to other organic compounds in other organisms through food chains.

43. At any point on the food chain an organism may use the carbon compounds in cell respiration to meet its energy needs. When this occurs, the carbon atoms are released as ________________.

44. The amount of carbon dioxide in the air is being greatly increased by the burning of fossil ________________.

The Phosphorus Cycle

45. The reservoir of the element phosphorus exists in various rock and soil ________________.

46. Plants absorb phosphate from the ________________ or ________________ solution.

47. Phosphate is first assimilated into organic compounds as ________________ phosphate and then passed to various other organisms through the food chain.

48. At any point on the food chain, an organism may use the organic compound containing phosphorus in respiration. When this occurs, the phosphate is released back to the environment in ________________.

49. What condition(s) are necessary for phosphorus to be recycled in the same ecosystem? ____

The Nitrogen Cycle

50. The main reservoir of nitrogen is N_2 in the air which is about ________ percent nitrogen gas.

51. Is it (True, False) that plants cannot assimilate nitrogen gas from the atmosphere?

52. To be assimilated by higher plants, nitrogen must be present as ________________ or ________________.

53. A number of bacteria and certain blue green algae can convert nitrogen gas to the ammonium form, a process called nitrogen ________________.

54. The bacteria have a mutualistic relationship with the ________________ family of plants.

55. Once fixed, nitrogen may be passed down food chains, excreted as ________________ and recycled as the mineral nutrient called ________________.

56. Nitrogen eventually returns back to the atmosphere because certain bacteria convert the nitrate and ammonia compounds back to ________________.

57. Is it (True, False) that all natural terrestrial ecosystems are dependent on the presence of nitrogen fixing organisms and legumes?

58. The human agricultural system bypasses nitrogen fixation by using fertilizers containing ______________ or ________________.

ENERGY FLOW

59. *Ecosystems run on solar energy which is exceedingly abundant, nonpolluting, constant, and everlasting* is the (First, Second, Third) basic principle of ecosystem function.

60. Identify four benefits derived from solar energy.

a. __________________, b. __________________

c. __________________, d. __________________

FLOW OF ENERGY AND DECREASING BIOMASS AT HIGHER TROPHIC LEVELS

61. *Large biomasses cannot be supported at the end of long food chains. Increasing population means moving closer on the food chain to the source of production* is the (First, Second, Third) ecological principle.

62. Biomass (Increases, Decreases) at each successive trophic level.

63. The energy available to producers is (Greater, Less) than that available to herbivores, therefore, the biomass of producers will be (Greater, Less) than that of herbivores.

64. The energy available to herbivores is (Greater, Less) than that available to carnivores, therefore, the biomass of herbivores will be (Greater, Less) than that of carnivores.

65. Indicate whether the following terms pertain to the first [1], second [2], or third [3] principle of ecosystem function.

[ ] biomass, [ ] recycling, [ ] solar energy, [ ] nonpolluting, [ ] carbon,

[ ] food chains, [ ] ammonium and nitrates

IMPLICATIONS FOR HUMANS

66. For each of the following statements regarding specific human actions toward the ecosystem, indicate whether the action is a violation of the first [1], second [2], or third [3] principle of ecosystem function.

[ ] lack of recycling
[ ] excessive use of fossil fuels
[ ] feeding largely on the third trophic level
[ ] excessive use of fertilizers
[ ] use of nuclear power
[ ] using agricultural land to produce meat

KEY WORDS AND CONCEPTS

matter	atom	element	compound
carbon dioxide	oxygen	carbon	nitrogen
phosphorus	hydrogen	air	electrons
protons	neutrons	neutral atom	charged atom
atomic number	atomic weight	isotope	covalent bond
ionic bond	orbital	molecule	ion
stability	abiotic	biotic	organic
mass	water	potential energy	kinetic energy
calorie	inorganic	energy	fossil fuels
chemical energy	organic molecules	photosynthesis	malnutrition
biomass	cell respiration	anaerobic respiration	urine
nutrient cycling	carbon cycle	phosphorus cycle	nitrogen cycle
legume plants	chlorophyll	cellulose	
nitrogen fixation			
Law of Conservation of Matter	First Law of Thermodynamics	Second Law of Thermodynamics	

SELF TEST

Circle the correct answer to each question.

1. Which of the following is not an important element in the structure of living organisms?

   a. carbon, b. aluminum, c. hydrogen, d. phosphorus

2. Organic molecules always contain

   a. nitrates, b. carbon-carbon bonds, c. minerals, d. water

3. Which of the following is not characteristic of matter?

   a. cannot be destroyed, b. can be changed, c. cannot be created
   d. can be rearranged

4. Energy can best be defined as:

   a. a nutrient, b. a vitamin, c. ability to do work, d. food

5. Which of the following is not a form of energy?

   a. protons, b. heat, c. light, d. motion

6. Which of the following is not characteristic of energy?

   a. cannot be destroyed, b. can be converted, c. cannot be created, d. can be recycled

7. In every energy conversion

   a. some energy is converted to heat, b. heat energy is lost,
   c. lost heat may be recaptured and reused, d. a and b

8. Which of the following is not a form of potential energy?

a. wood, b. light, c. coal, d. oil

9. Green plants obtain energy from

a. light, b. soil, c. nutrients, d. water

10. For the plant, the most important product of photosynthesis is

a. carbon dioxide, b. glucose, c. oxygen, d. chlorophyll

11. A waste or by-product of photosynthesis is

a. oxygen, b. carbon dioxide, c. water, d. glucose

12. Consumers must feed on preexisting organic material to obtain

a. nutrients, b. energy, c. both a and b,
d. neither a or b - consumers make their own nutrients and energy

13. The chemical process through which food is broken down to release energy is known as

a. cell metabolism, b. photosynthesis, c. cell respiration, d. digestion

14. Which of the following is not recycled in natural ecosystems

a. nitrogen, b. carbon, c. energy, d. phosphorus

15. The carbon atoms in the food you eat will

a. be exhaled as carbon dioxide, b. become part of your body tissues,
c. pass through as fecal wastes, d. a, b, and c

16. In the phosphorus cycle, soil phosphate first enters the ecosystem through

a. plants, b. herbivores, c. carnivores, d. decomposers

17. Nitrogen fixation refers to

a. releasing nitrogen into the air, b. converting it to a chemical form plants can use,
c. repairing broken molecules, d. applying fertilizer

18. The nitrogen cycle relies on

a. mutualism, b. commensalism, c. parasitism, d. synergism

19. By far, the largest biomass is found in

a. producers, b. consumers, c. decomposers, d. humans

20. Humans generally ignore

a. recycling, b. solar energy, c. their population growth, d. a, b, and c

ANSWERS TO STUDY QUESTIONS

APPENDIX C

1. solid, liquids, gas; 2. matter; 3. atoms; 4. element; 5. atoms; 6. compound; 7. different; 8. 96; 9. E,C,C,E,C,E,E,C,C,E; 10. -,-,-,+,+; 11. protons, neutrons, electrons; 12. protons; 13. protons; 14. must be different, carbon, oxygen; 15. will be the same, isotopes; 16. covalent, ionic; 17. orbitals, specific; 18. two, eight; 19. true; 20. completely, empty; 21. same; 22. bonding together; 23. covalent; 24. molecule; 25. carbon; 26. hydrogen, oxygen, nitrogen, phosphorus, sulfur; 27. covalent; 28. gaining, losing; 29. ion; 30. negative; 31. negative; 32. attract; 33. attraction of opposite charges; 34. different charges are required; 35. rocks, minerals; 36. a; 37. different; 38. greater; 39. energy; 40. releases; 41. has greater stability as water; 42. true, photosynthesis; 43. energy

ELEMENTS, LIFE, ORGANIZATION, AND ENERGY

1. basically the same, the same; 2. same; 3. N, CHOPS; 4. simple, extremely complex; 5. 3, oxygen, nitrogen, carbon dioxide; 6. hydrogen, oxygen; 7. (carbon dioxide, water), (water, water), (oxygen gas, air), (nitrogen gas, air), (phosphate ion, dissolved in water or in rock or soil minerals); 8. +, -, -, +; 9. space, mass; 10. solids, liquids, gas; 11. ability to do work; 12. light, radiation, heat; 13. -, -, +, +, -, -, +, +, +; 14. kinetic; 15. potential; 16. chemical; 17. calorie; 18. +, +; 19. +, +; 20. L, H, H, L, H, H; 21. photosynthesis; 22. chlorophyll; 23. 5; 24. glucose; 25. materials, energy; 26. materials, energy; 27. respiration; 28. reverse; 29. P, B, R, P, B, P, B, B; 30. all are marked; 31. malnutrition; 32. malnutrition; 33. calorie, nutrients; 34. fecal; 35. detritus; 36. cellulose; 37. b and c; 38. first; 39. recycled; 40. recycled; 41. carbon dioxide, air; 42. photosynthesis; 43. carbon dioxide; 44. fuels; 45. minerals; 46. soil, water; 47. organic; 48. urine; 49. must be deposited in that ecosystem; 50. 78; 51. true; 52. ammonium, nitrates; 53. fixation; 54. legume; 55. ammonium, nitrates; 56. nitrogen gas; 57. true; 58. ammonium, nitrate; 59. second; 60. abundant, nonpolluting, constant, everlasting; 61. third; 62. decreases; 63. greater, greater; 64. greater, greater; 65. 3, 1, 2, 2, 1, 3, 1; 66. 1, 2, 3, 1, 2, 3

ANSWERS TO SELF TEST

1. b; 2. b; 3. b; 4. c; 5. a; 6. d; 7. d; 8. b; 9. a; 10. b; 11. a; 12. c; 13. c; 14. c; 15. d; 16. a; 17. b; 18. a; 19. a; 20. d

CHAPTER 3

ECOSYSTEMS: WHAT KEEPS THEM THE SAME? WHAT MAKES THEM DIFFERENT

We have seen that every ecosystem consists of a grouping of plants, animals, and microbes in various roles as producers, consumers, and decomposers being fed upon or feeding upon each other in a complex food web. Overall, we see that the ecosystem is sustained by a cycling of nutrients within the system and a flow of energy through the system. But, what prevents herbivores from eating up all the producers? What prevents carnivores from devouring all their prey? What prevents parasites from killing off the entirety of their host population? What prevents one species from overpopulating and crowding all others? It should be obvious that if any of these things did happen, it would destroy the ecosystem. But this does not answer the question of why these events do not happen in natural ecosystems.

Note that we said these events do not *generally* happen! However, when we view humans in the context of the total biosphere, we do see some of these things occurring. Humans are overpopulating and along with their developments, are destroying and crowding out and destroying many species of wildlife and even whole ecosystems. Also humans are such effective predators that we have hunted and are continuing to hunt many species of wildlife into extinction. Further damage is being done by pollution, and by introducing species from one ecosystem into another. These actions, if allowed to continue, can eventually lead to the collapse of the global ecosystem, the biosphere, and, of course, human civilization along with it. Sustaining the biosphere and the human ecosystem within it will depend, first, on our understanding of the checks and balances which underlie the maintenance of natural ecosystems and, second, on our applying such checks and balances to our own ecosystem.

STUDY QUESTIONS

THE KEY IS BALANCE

1. Is it (True, False) that there are natural laws and rigid structures that prevent ecosystems from changing?

2. The only thing that maintains natural ecosystems over long periods of time is that all the dynamic relationships are in ____________________.

ECOSYSTEM BALANCE IS POPULATION BALANCE

3. Each species found in an ecosystem is represented by a larger, interbreeding, reproducing group called a ____________________.

4. If an ecosystem is to remain stable over a long period of time, the size of each population must remain relatively ____________________.

5. In order for a population to remain constant, its reproductive rate must be balanced by and equal ____________________ rate.

Biotic Potential Versus Environmental Resistance

6. All the aspects of a species that lend to its capacity to increase its population size are referred to as its ____________________ potential.

7. Which of the following is **NOT** an example of biotic potential?

   a. reproduction, b. disease, c. recruitment, d. defense mechanisms

8. All the factors that cause the dieoff of individuals in the population are categorized as ____________________ resistance.

9. Which of the following is **NOT** an example of environmental resistance?

   a. reproduction, b. disease, c. lack of food, d. lack of shelter

10. If biotic potential is higher than environmental resistance, the population will (Increase, Decrease).

11. Population explosions occur when biotic potential is (Higher, Lower) than environmental resistance.

12. When a population is more or less stable in terms of numbers, biotic potential is (Higher Than, Lower Than, Equal To) environmental resistance.

Density Dependence and Critical Numbers

13. The number of individuals per unit area is termed ________________.

14. For many species, it has been found that if a population drops below a certain number, it cannot recover even when conditions are favorable. Such a number is referred to as the ____________________.

15. If a population drops below a certain critical number, ________________ is virtually assured.

MECHANISMS OF POPULATION BALANCE

16. Is it (True, False) that the dynamic balances which exist between biotic potential and environmental resistance may be easily upset by humans?

17. Is it (True, False) that when balances are upset, there are population explosions of some species and extinctions of other species?

Predator-Prey and Host-Parasite Balances

18. Which of the following organisms can be an important factor in limiting the growth of a population?

    a. a large carnivore, b. a viral disease organism, c. a parasitic organism, d. all of these

19. Which of the following organisms may have their populations limited by various parasites?

    a. plants, b. herbivores, c. bacteria, d. all of these

20. All the various organisms that adversely affect a certain species may be referred to as the species' ____________________ enemies.

21. If a herbivore population is not controlled by natural enemies, it is likely that the herbivore population will (Increase, Decrease) and the vegetation it feeds on will (Increase, Decrease).

22. Define the **principle of ecosystem stability.**

____

23. Is it (True, False) that a natural enemy can be so effective that it causes the extinction of its prey or host? If you said "Yes", give an example ____________________

24. If individuals are stressed by suboptimal conditions, indicate whether they would [+] or would not [-] exhibit the following characteristics.

[ ] Less able to compete successfully with other species.
[ ] More vulnerable to attack by predators.
[ ] More vulnerable to attack by parasitic and disease organisms.
[ ] More likely to attempt to disperse from the area.

Specialization to Habitats and Niches

25. Indicate whether the following conditions would [+] or would not [-] prevent a population from growing and spreading into adjacent areas.

[ ] It cannot outcompete species that are already there.
[ ] It is less well adapted to conditions in the adjacent area.

26. Where an organism lives is called its ____________________.

27. Where, when, and what an organism feeds on is called its ____________________.

28. Is it (True, False) that all the different species of animals in an ecosystem are inevitably competing with each other?

29. Competition among various species of animals in an ecosystem is minimal because:

a. they feed on different things. (True, False)
b. they feed in different locations. (True, False)
c. they feed at different times. (True, False)

30. Indicate whether the following "introductions" have [+] or have not [-] upset or effectively destroyed ecosystems.

[ ] Introducing plant species that outcompete native species.
[ ] Introducing herbivores that outcompete native grazing animals.
[ ] Introducing carnivores that outcompete native carnivores.
[ ] Introducing carnivores that prey species do not recognize.

Competition Among Plants and Plant-Herbivore Balance

31. A factor that plays an important role in maintaining diversity in a plant community is ____________________.

32. If a plant species is in abundance, herbivory by insect pests will (Increase, Decrease).

33. The most sustainable plant community is a (Diverse, Monoculture) system that includes (Numerous, Few) species all at a relatively (High, Low) density.

Fire

34. In which of the below ecosystems can fire be used as a management tool?

a. pine forests, b. hardwood forests, c. tropical rain forests, d. grasslands

35. Indicate whether the following events are [+] or are not [-] the results of controlled ground fires.

[ ] removal of dead litter
[ ] permanent harm of mature trees
[ ] seed release of some tree species
[ ] destruction of wood-boring insects
[ ] crown fires

Territoriality

36. Aggressive defense of habitat is called ____________________.

37. Is it (True, False) that territorial behavior by an animal results in priority use of those resource available in a specific area?

ECOSYSTEM CHANGE: SUCCESSION

38. The process whereby the growth of one community causes changes that make the environment more favorable to a second community and less favorable to the first is called ____________________.

Primary Succession

39. If an area has not been previously occupied, the process of initial invasion and then progression from one ecosystem to another is referred to as ____________________ succession.

40. In what order (1 to 4) would the following plant communities appear during primary succession?

[ ] trees, [ ] mosses, [ ] shrubs, [ ] larger plants

41. How do mosses change the conditions of the area so that it can support larger plants?

____

Secondary Succession

42. The redevelopment of an ecosystem in an area that previously supported the same kind of ecosystem is a process referred to as ____________________ succession.

43. Suppose that a section of eastern deciduous forest was cleared for agriculture and later abandoned. In what order (1 to 4) would the following species invade the area?

[ ] deciduous trees, [ ] crabgrass, [ ] pine trees, [ ] grasses

44. What prevents hardwood species of the deciduous forest from reinvading the area immediately? ______________________________

The Climax Ecosystem

45. When succession reaches a point at which all the species present continue to reproduce in proportion to each other and no further change occurs, the system is called the ______________ ecosystem.

46. Is it (True, False) that the dominant plant forms in all climax ecosystems are the same?

47. Is it (True, False) that the climax ecosystem will maintain itself regardless of changes in climatic, abiotic, and biotic factors?

Degree of Imbalance and Rate of Change: Succession, Upset, or Collapse

48. A very gradual and orderly change in an ecosystem is referred to as ______________.

49. A change which results in a population explosion of one species at the expense of others is called an ______________.

50. A change that results in the death of all or nearly all species is referred to as a ______________ of the ecosystem.

51. Human induced changes often result in ______________ or ______________.

IMPLICATIONS FOR HUMANS

52. The human population is exploding because we have effectively

a. increased our biotic potential. (True, False)
b. decreased our environmental resistance. (True, False)

53. List the first five of the ten ways in which humans are causing the upset of natural ecosystems.

a. ______________________________

b. ______________________________

c. ______________________________

d. ______________________________

e. ______________________________

54. Indicate which of the following human actions will have a positive [+] or negative [-] impact on ecosystem stability.

[ ] Controlling population size.
[ ] Decreasing the gross national product.
[ ] Becoming proponents of conservation legislation.
[ ] Supporting conservation organizations.
[ ] Gaining an understanding of conservation issues through education.

KEY WORDS AND CONCEPTS

population	dynamic balance	biotic potential	environmental resistance
recruitment	recruitment success	population explosion	critical number
extinction	overgrazing	natural enemies	niche
territoriality	fire resistant	ground fire	crown fire
succession	primary succession	secondary succession	climax ecosystem
ecological upset	ecological collapse	monoculture	

SELF TEST

1. All the aspects of a species that favor its capacity to increase its population size are referred to as

   a. biotic potential, b. fertility, c. environmental resistance, d. regenerative capacity

2. Which of the following is not an example of biotic potential?

   a. reproduction, b. disease, c. recruitment, d. defense mechanisms

3. All the factors that cause the dieoff of individuals in a population are categorized as

   a. biotic potential, b. fertility, c. environmental resistance, d. regenerative capacity

4. Which of the following is not an example of environmental resistance?

   a. reproduction, b. disease, c. lack of food, d. lack of shelter

5. If biotic potential and environmental resistance are equal, the population will

   a. increase, b. decrease, c. stay the same, d. there is no way of predicting

6. If biotic potential is greater than environmental resistance, the population will

   a. increase, b. decrease, c. stay the same, d. there is no way of predicting

7. If biotic potential is less than environmental resistance, the population could realize

   a. its threshold level.
   b. extinction.
   c. a temporary decrease in numbers.
   d. All of these are possibilities.

8. Predators and other natural enemies

a. limit the increase of herbivore populations.
b. prevent herbivores from overgrazing.
c. indirectly prevent erosion.
d. All of these are correct.

9. Ecosystem stability is best maintained through high species

a. numbers, b. densities, c. diversity, d. a and c

10. When rabbits were introduced into Australia, they over-populated and became a problem because

a. there were no carnivores in Australia.
b. there were no herbivore competitors in Australia.
c. the rabbits increased their biotic potential.
d. All of the above contributed to rabbit increase.

11. In which of the following ecosystems can fire not be used as a management tool?

a. grasslands, b. coniferous forests, c. hardwood forests, d. a and b

12. Which of the following is not an intended function of ground fires?

a. removal of dead litter
b. seed release of some tree species
c. starting crown fires
d. destruction of wood-boring insects

13. Competition for feeding niches between organisms is minimized by feeding

a. on different things.
b. at different times.
c. at different locations.
d. All of these reduce competition.

14. Which of the following would be the last plant community to appear during primary succession?

a. trees, b. mosses, c. shrubs, d. grasses

15. Which of the following would be the first plant community to appear during secondary succession?

a. deciduous trees, b. crabgrass, c. pine trees, d. grasses

16. Aggressive defense of habitat is called

a. biotic potential, b. succession, c. territoriality, d. the niche

17. The process whereby the growth of one community causes changes that make the environment more favorable to a second community and less favorable to the first is called

a. biotic potential, b. succession, c. territoriality, d. the niche

18. The redevelopment of an ecosystem in an area that previously supported the same kind of ecosystem is referred to as

a. primary succession
b. the climax community
c. a monoculture
d. secondary succession

19. Which of the following is not a characteristic of climax ecosystems?

a. have maximum species diversity
b. have different dominant plant forms
c. are self-perpetuating
d. will maintain themselves regardless of climatic, abiotic, or biotic changes

20. Which of the following human actions will have a negative impact on ecosystem stability?

a. increasing population size
b. decreasing the gross national product
c. supporting conservation organizations
d. understanding conservation issues

ANSWERS TO STUDY QUESTIONS

1. true; 2. balance; 3. population; 4. stable; 5. death; 6. biotic; 7. b; 8. environmental; 9. a; 10. increase; 11. higher; 12. equal to; 13. density; 14. threshold; 15. extinction; 16. true; 17. true; 18. d; 19. d; 20. natural; 21. decrease, increase; 22. "Species diversity provides ecosystem diversity"; 23. true, humans; 24. all +; 25. all +; 26. habitat; 27. niche; 28. false; 29. all true; 30. all +; 31. herbivory; 32. increase; 33. diverse, numerous, low; 34. b and c; 35. +, +, +, +, -; 36. territoriality; 37. true; 38. succession; 39. primary; 40. 4, 1, 3, 2; 41. provide first layer of humus; 42. secondary; 43. 4, 1, 3, 2; 44. temperature too hot; 45. climax; 46. false; 47. false; 48. succession; 49. upset; 50. collapse; 51. upsets, collapse; 52. true, true; 53. destruction of natural ecosystems, diverting and damming waterways, discharging pollutants, introduction of exotics, overgrazing; 54. all +

ANSWERS TO SELF TEST

1. a; 2. b; 3. c; 4. a; 5. c; 6. a; 7. d; 8. d; 9. d; 10. c; 11. d; 12. c; 13. d; 14. a; 15. b; 16. c; 17. b; 18. d; 19. d; 20. a

CHAPTER 4

ECOSYSTEMS: ADAPTATION AND CHANGE OR EXTINCTION

At this point in your study of how ecosystems work, you should be becoming aware of the fact that even though organisms attempt to maintain a dynamic balance between biotic potential and environmental resistance - the struggle is everlasting. This is the case because ecosystems are in a constant state of change. These changes are the result of subtle or dramatic alterations in the abiotic or biotic characteristics of the ecosystem. However, whether the change is subtle (e.g., slight change in pH) or dramatic (e.g., turning a grassland into a corn field), each species has only three response alternatives to these changes. These alternatives are to adapt (adaptation), move (migration), or die (extinction). The choice of which alternative is used will be based on the information available, or not, in the gene pool of a species.

Chapter 4 provides an overview of how genetics plays the major role in adaptation, change, or extinction. As soon as humans realized how genetics plays this role, we developed technologies that allowed us to manipulate (e.g., hybridization and genetic engineering) the gene pools of selected species. We now realize that these manipulations of gene pools also required the development of safeguards against natural selection. For example, hybrid seed corn was developed to produce lots of corn, not to combat environmental resistance. In order to keep hybrid seed corn alive and to combat environmental resistance, farmers need to provide tremendous amounts of water, energy, and nutrients. Further examples of how the gene pool of a species hinders human efforts to control environmental resistance will be discussed in later chapters on pest control.

STUDY QUESTIONS

HOW SPECIES CHANGE AND ADAPT ... OR DON'T

SOME BASIC GENETICS

Traits and Genes

1. A key observable feature of all species is that there is ______________ among individuals.

2. The term ______________ refers to any characteristic of an organism.

3. Provide an example of a trait related to:

   a. physical appearance. ________________
   b. tolerance. ______________
   c. behavior. ______________

4. The underlying basis of traits is hereditary or ____________.

5. The hereditary component of all traits of all organisms is encoded in ____________ molecules.

6. These DNA molecules are the organism's ______________.

7. The collection of all the DNA molecules or genes within a cell of an organism is called its ________________ makeup.

Sexual Reproduction

8. The genetic makeup of nearly all organisms consists of (One, Two, Three, Four) complete set(s) of genes or (One, Two, Three, Four) pair(s).

9. Each individual gene of a gene pair is called an ________________.

10. Indicate whether each of the below cell types would have one [1] or two [2] alleles of a gene pair.

    [ ] body cells, [ ] sperm cell, [ ] egg (ovum), [ ] fertilized egg

11. A sperm cell entering an egg is called the process of ________________.

12. Each fertilization will result in a different combination of ________________ and this combination will be different from that of either parent.

13. Each member of a population (except for identical twins) has the/a (Same, Different) genetic makeup?

Mutations

14. Random "accidental" changes in the DNA are called ________________.

15. Is it (True, False) that mutations are another source of genetic variation within populations?

16. Is it (True, False) that mutations can be inherited or passed on from one generation to another?

17. Identify three abiotic factors that can cause mutations.

    a. ________________, b. ________________, c. ________________

18. Genetic variation among individuals of a population is inevitable through segregation and recombination of genes in sexual ________________ and the occurrence of ________________.

19. The genetic makeup of **cloned** organisms is (Identical, Different) barring mutations.

GENE POOLS AND THEIR CHANGE

20. Each individual has (One, Two, Three, Four) complete sets of genes and (One, Two, Three, Four) different alleles of any particular gene.

21. Is it (True, False) that there can be only two sets of genes but many different alleles of a gene?

22. The ________________ pool of a species is the total of all the different alleles of each gene that exist in the entire population of the species.

23. Is it (True, False) that some alleles are carried by large numbers of individuals within the population and other alleles are carried by only a few individuals?

24. If an individual that carries the "big A" allele reproduces more than one carrying a "little a" allele, the "big A" allele will become (More, Less) abundant in the population.

25. Any new allele that inhibits ________________ of an individual will be weeded out of the population.

Change Through Selective Breeding

26. Breeding plants or animals in order to have the offspring express a particular trait more than the parents is called ________________ ________________.

27. Selective breeding requires a (Repetitive, One-time) process in order to develop the desired trait.

28. Selective breeding (a. Increases, b. Decreases, c. both a and b) certain alleles in the gene pool of a population.

29. All the different breeds of dogs are derived from the (Same, Different) wild dog gene pool.

30. Cross breeding with a different but closely related species is called ________________.

31. Offspring that result from cross breeding different but closely related species are called ________________.

32. Selective breeding of hybrids selects for ________________ traits and against ________________ traits.

33. The process of isolating specific genes from one species and introducing these genes directly into another species is called genetic ________________.

34. Is it (True, False) that the ultimate results of modifying gene pools through genetic engineering are entirely predictable?

Change Through Natural Selection

35. The numbers of individuals within a population is determined by the balance between ________________ potential and ________________ resistance.

36. Biotic potential tends to (Increase, Decrease) the number of individuals within a population.

37. Environmental resistance tends to (Increase, Decrease) the number individuals within a population.

38. Is it (True, False) that most species reproduce more offspring than can possible survive?

39. In nature, every generation of every species is subjected to an intense selection of ________________ and ________________.

40. New alleles that provide for or enhance a trait that aids in survival and reproduction will (Increase, Decrease) in the gene pool of the population.

41. ________________ selection is the process in nature that determines which alleles enhance or detract from survival and reproduction.

42. Indicate whether the following adaptations would [+] or would not [-] support the survival and reproduction of organisms.

[ ] Coping with climatic and other abiotic factors.
[ ] Obtaining food and water.
[ ] Escaping from predation.
[ ] Inability to attract mates.
[ ] Ability to migrate to adjacent areas.

43. Is it (True, False) that just because an organism survives, it will automatically also reproduce?

Specialization to Niches and Habitats

44. Is it (True, False) that a species can become **too** specialized to the particular habitat and niche in which it exists?

45. For example, is it (True, False) that the necks of giraffes could get **too** long?

Speciation

46. If a species migrates or disperses to a new location where biotic and/or abiotic conditions are somewhat different, which of the following events could happen?

a. The migrant population may die out. (True, False)
b. Some individuals may survive and reproduce. (True, False)
c. There will be a strong selection pressure for alleles that enhance survival. (True, False)
d. Alleles in the gene pool of the migrant population will become different from those in the gene pool of the original population. (True, False)
e. Speciation could occur. (True, False)

47. List two examples of modification and speciation occurring in nature.

a. ____

b. ____

48. In describing the concept of "survival of the fittest", the "fittest" are those individuals with variations of traits that best enable their ______________ and ______________.

DEVELOPMENT OF ECOSYSTEMS

49. Every species exists and can only exist in the context of:

a. producers, b. consumers, c. decomposers, d. the whole ecosystem

50. If the gene pool of a prey species contains an allele providing a new trait to better escape predators, then what needs to occur in the gene pool of the predator species in order to maintain a balance?

a. nothing
b. selection for alleles providing traits for more effective capture of prey

51. If there is a lasting imbalance between two species, what will happen to the species on the down-side?

a. nothing, b. become extinct

52. If the process of **selection** is a process of modifying gene pools of existing species, then:

a. What exists can be modified. (True, False)
b. What does not exist can be created. (True, False)
c. How an ecosystem develops depends on what species are present at the beginning of the process. (True, False)

53. Identify the prime herbivore on:

a. The Galapagos Islands ________________
b. Australia ____________________

54. Why are the prime herbivores on the Galapagos Islands and Australia different species?

____

55. Is it (True, False) that if you introduced the Galapagos turtles into Australia, they would become the prime herbivores?

56. Is it (True, False) that if you introduced cattle into Australia, they would become the prime herbivores?

EVOLUTIONARY SUCCESSION

57. List the three options available to each species that is ill-adapted to a new ecosystem.

a. ________________, b. __________________, c. __________________

LIMITS OF CHANGE: MAKING IT OR NOT MAKING IT

58. Indicate whether the following are criteria for [+] or not [-] making it as a species.

[ ] High degree of genetic diversity.
[ ] Rapid abiotic/biotic changes.
[ ] Wide geographic distribution.
[ ] High reproductive capacity.

IMPLICATIONS FOR HUMANS

59. Indicate whether the following conditions in the human ecosystem are criteria for [+] or not [-] making it as a species.

[ ] Causing rapid abiotic/biotic changes.
[ ] Low reproductive rate.
[ ] Decreasing genetic diversity in plants and animals.
[ ] Developing a sustainable, balanced, human ecosystem.

KEY VOCABULARY OR CONCEPTS

traits	genes	DNA	genetic makeup
alleles	fertilization	mutations	cloning
gene pool	selective breeding	hybrid	hybridization
genetic engineering	natural selection	adaptations	speciation
modification	"fittest"	species balance	species imbalance
evolutionary succession	migration	adaptation	extinction
genetic diversity	geographic distribution	reproductive capacity	major event

SELF TEST

Circle the correct answer to each question.

1. An observable feature of all species is that their traits vary in terms of

   a. physical appearance, b. tolerance, c. behavior, d. all of these traits

2. Which of the following statements is not true concerning DNA molecules? DNA molecules are

   a. the site of genetic information within organisms.
   b. found only in sperm or egg cells.
   c. capable of being influenced by genetic engineering.
   d. one of the locations for mutations to occur.

3. How many alleles of a gene pair would normally be found in a sperm cell?

   a. one, b. two, c. three, d. four

4. How many alleles of a gene pair would normally be found in a skin cell?

   a. one, b. two, c. three, d. four

5. A sperm cell entering an egg cell is called the process of

   a. mutation, b. variation, c. fertilization, d. reproduction

6. Random accidental changes in DNA are called

   a. mutations, b. variations, c. fertilization, d. reproduction

7. Which of the following is not known to cause mutations?

   a. chemicals, b. hybridization, c. drugs, d. radiation

8. The total of all the different alleles of each gene that exists in the entire population of a species is called the

   a. allele frequency, b. mutation rate, c. gene pool, d. variation index

9. Which of the following statements is not correct regarding selective breeding?

a. It is a one-time event.
b. It increases the frequency of certain alleles in the gene pool of a population.
c. It decreases the frequency of certain alleles in the gene pool of a population.
d. It only works with individuals of the same species.

10. Cross breeding with a different but closely related species is called

a. selective breeding, b. hybridization, c. genetic engineering, d. cloning

11. The process of isolating specific genes from one species and introducing these genes directly into another species is called

a. selective breeding, b. hybridization, c. genetic engineering, d. cloning

12. Every generation of every species is subject to an intense selection of

a. survival, b. reproduction, c. migration, d. a and b

13. An organism well adapted for survival would not have which of the following characteristics?

a. wide geographic distribution
b. lots of offspring
c. a narrow range of tolerance
d. ability to attract mates

14. If a species migrates to a new location where biotic and/or abiotic conditions are somewhat different, which of the following events could happen?

a. The migrant population could die out.
b. Some individuals may survive and reproduce.
c. Speciation could occur.
d. All of these events are possible.

15. Which of the following statements concerning gene pool selection is false?

a. What exists can be modified.
b. What does not exist can be created.
c. How an ecosystem develops depends on the species present in the beginning.
d. All of the above are correct concerning natural selection.

ANSWERS TO STUDY QUESTIONS

1. variation; 2. trait; 3. the way an organism looks, susceptibility to diseases, ability to accomplish various tasks; 4. genetics; 5. DNA; 6. genes; 7. genetic; 8. two, one; 9. allele; 10. 2, 1, 1, 2; 11. fertilization; 12. genes; 13. different; 14. mutations; 15. true; 16. true; 17. radiation, drugs, chemicals; 18. reproduction, mutations; 19. identical; 20. two, two; 21. true; 22. gene; 23. true; 24. more; 25. reproduction; 26. selective; 27. repetitive; 28. c; 29. same; 30. hybridization; 31. hybrids; 32. desired, undesired; 33. engineering; 34. false; 35. biotic, environmental; 36. increase; 37. decrease; 38. true; 39. survival, reproduction; 40. increase; 41. natural; 42. all +; 43. false; 44. true; 45. true; 46. all true; 47. plant growth in alpine tundra, arctic fox characteristics; 48. survival, reproduction; 49. d; 50. b; 51. b; 52. true, false, true; 53. turtles, kangaroos; 54. started with different herbivore species; 55. false; 56. true; 57. adaptation, migration, extinction; 58. +, -, +, +; 59. -, +, -, +

ANSWERS TO SELF TEST

1. d; 2. b; 3. a; 4. b; 5. c; 6. a; 7. b; 8. c; 9. a; 10. b; 11. c; 12. d; 13. c; 14. d; 15. b

CHAPTER 5

THE POPULATION PROBLEM:
ITS DIMENSIONS AND CAUSES

I cannot believe that the principal objective of humanity is to establish experimentally how many human beings the planet can just barely sustain. But I can imagine a remarkable world in which a limited population can live in abundance, free to explore the full extent of man's imagination and spirit.

Philip Handler, Past President
National Academy of Sciences

Can we not all see the common sense and taste the appeal of Handler's statement? Yet from observations of current trends one could well be led to the conclusion that humans are, indeed, embarked on a great global experiment to test just how many humans the Earth can just barely sustain.

Over the past several decades, the human population has been growing at a rate of 70-90 million people per year. This means that an additional number of people equivalent to about a third of the United State's population is coming on board Spaceship Earth each year, and this trend is continuing with little abatement. There may be virtue in additional numbers if people can live in abundance, free to explore their full potential. Unfortunately, most of these new arrivals are boarding in poor, less developed countries in situations of poverty with very limited opportunities. But this is only half the tragedy; the other half is that the Earth's resources are being depleted, undercut and destroyed in order to sustain these new arrivals at all, a trend that, if continued, can only lead to upset and collapse of the entire biosphere with unprecedented suffering for all.

At what point may the collapse occur? This will depend on the degree of our ability to exercise ecological management of resources. With current poor ecological management of resources, particularly pollution and depletion of water, it is highly doubtful that the Earth can sustain even the present population. But, with sound ecological management of resources, the Earth may be able to support up to double the present population. Either way, it is imperative to address the population problem.

Study shows that it is not incidental that high birth rates and, hence, population growth are connected with poverty. Poverty and high birth rates are entwined in a vicious cycle. To break this cycle, we must understand it and attack it from the side of poverty as well as birth rate itself. Conveying this understanding and presenting the avenues of attack are objectives of this section.

STUDY QUESTIONS

DIMENSIONS OF THE POPULATION PROBLEM

THE EXPLODING HUMAN POPULATION

1. From its origin up until the 1700s the human population hovered around a few hundred ________________.

2. In the period from the 1700s through the 1800s the human population changed from a condition of ________________ growth to a condition of ________________ growth.

3. a. It took from the time of human origin until about 18 ________ for the human population to reach one billion.
   b. The second billion was added ________ years later in 19 ________.
   c. The third billion was added ________ years later in 19 ________.
   d. The fourth billion was added ________ years later in 19 ________.
   e. The fifth billion was added ________ years later in 19 ________.
   f. The sixth billion will probably be added in ________ years in 19 ________.

4. This phenomenon of increasingly rapid growth is called a population ______________.

5. What is happening to the Earth's essential resources in the process?
   ____

6. Most of the new arrivals are being born into conditions of ______________ disparity between nations.

7. Rather than being concerned with the maximum number of people that the Earth can support, we should be more concerned with the ________________ regard of lifestyles.

RICH AND POOR NATIONS

8. The countries of the world, excluding Russia and East European nations, fall into three major economic categories. Identify these categories with examples of countries or regions that are in each category. (see Fig. 5-1)

Category	Countries\Regions
a. ________________________	United States, Japan, Western Europe
b. ________________________	Latin America, East Asia
c. ________________________	East and Central Africa, India, China

9. Highly developed nations have about ________ percent of the world's population, but they control about ________ percent of the world's wealth.

10. Developing countries have ________ percent of the world's population, but control only ________ percent of the world's wealth.

11. Between ________ percent and ________ percent of people in highly developed countries are recognized as (Rich, Middle Income, Poor).

12. Define what it means to be **poor** ____

13. World wide, the primary economic concern of at least a (Thousand, Million, Billion) people is simply day-to-day survival.

POPULATION AND POVERTY

14. The population explosion is most intense in:

   a. highly developed countries, b. moderately developed countries, c. less developed countries

15. Less developed countries will double their present populations in ________ to ________ years.

16. Populations in highly developed countries are approaching ____________.

Fertility and Population Profiles

17. The average number of children born to each woman in her lifetime is defined as ______________ rate.

18. Indicate whether the following fertility rates will lead to an increase [+], decrease [-], or stability [0] in population growth.

[ ] Fertility rate < 2
[ ] Fertility rate = 2
[ ] Fertility rate > 2

19. Which of the above fertility rates is known as **replacement fertility?** __________

20. The age makeup of a population can be demonstrated graphically as a population ______________.

21. Match the below shapes of population profiles with the fertility rates given in question #18.

a. Column-shaped: Fertility rate = ________
b. Pyramid-shaped: Fertility rate = ________
c. Upside down pyramid: Fertility rate = ________

22. If fertility rates of less developed countries remain relatively constant over the next five years, one can predict that in five years their population pyramid will look like:

a. straight column, b. pyramid with a broader base, c. upside down pyramid

Growth Momentum

23. Even if the fertility rates of populations in less developed countries dropped to replacement immediately, their population size would continue to increase because of growth ______________.

24. Given the growth momentum of less developed countries, their populations will continue to grow markedly until 20 ______ .

25. In the next 50 years, developed countries will probably represent only ______ percent of the world's population.

26. This will lead to the rich getting ______________ and the poor getting ______________.

Population Growth Undercuts Economic Gain and the Debt Crisis

27. The **gross national product per capita** is a measure of the average standard of ______________ in a country.

28. Is it (True, False) that the standard of living in countries with low populations will be higher when compared to countries with high populations.

29. Give the formula for **real** economic growth.

real economic growth = ____________________ - ____________________

30. Countries faced with a **debt crisis**:

a. pay more than half of all their earnings in interest. (True, False)
b. are unlikely to obtain additional credit from lending agencies. (True, False)
c. are faced with high unemployment. (True, False)
d. are faced with inflation rates topping 100 percent per year. (True, False)
e. are faced with deteriorating public services. (True, False)

POPULATION, POVERTY, AND THE ENVIRONMENT

31. Indicate which of the following environmental problems are [+] or not [-] the result of growing populations in less developed countries.

[ ] overgrazing, [ ] overcultivating, [ ] overcutting of forests

32. Is it (True, False) that those of us living in highly developed countries should **not** be concerned about environmental problems listed in question #31.

THE POPULATION EXPLOSION: ITS CAUSE AND POTENTIAL SOLUTION

33. List six factors that affect births and deaths.

a. ____________________, b. ____________________, c. ____________________

d. ____________________, e. ____________________, f. ____________________

34. Increase (or decrease) in population, excluding immigration or emigration, is given by the number of ____________________ minus number of ____________________ .

BIRTH RATES, DEATH RATES AND THE POPULATION EQUATION

35. The number of births per 1000 people per year is referred as the ________________ birth rate.

36. Similarly, the crude death rate refers to the number of ____________________ per (10, 100, 1,000) per year.

37. Number of births minus the number of deaths is referred to as the ________________ increase (or decrease).

38. Subtracting the crude death rate from the crude birth rate (CBR-CDR) and then dividing by 10 gives the rate of change in ____________________.

39. Are the growth rates of developed countries (Higher, Lower) than the growth rates in low and moderate income nations (see Table 5.1)?

40. Doubling time for a population may be found by dividing the percent rate of increase into (7, 70, 700).

41. The goal of each country should be to have a (Long, Short) doubling time.

CAUSE OF THE POPULATION EXPLOSION

42. Prior to the 1800's throughout the world, CBRs were in the order of ________________ and CDRs were almost equal. Therefore, population growth rates were (Slow, Fast).

43. Prior to the 1800s it was common for parents to have 7-10 children, but it was also common to have several of these children ________________ in infancy or early childhood.

Reduction in Infant and Childhood Mortality

44. Beginning in the 1800s, mortality began to be reduced by what factors?

    a. ______________________________

    b. ______________________________

45. Most profound was a dramatic reduction in the deaths occurring among ________________ .

46. This reduction in infant and childhood mortality can be referred to as an increase in ______________________.

Change from Prereproductive to Postreproductive Death

47. Prereproductive death means that death occurs ________________ having children, whereas death after bearing children is called ________________ death.

48. (Pre-, Post-) reproductive death has greater effect on preventing a population from growing.

The Minor Impact of Postreproductive Longevity

49. Is it (True, False) that extending life from 60 to 80 will have a significant impact on population growth?

50. Is it (True, False) that decreasing longevity from 60 to 40 will have a lasting impact on reducing the rate of population growth, again assuming birthrate remains constant?

The Minor Impact of Accidents and Natural Disasters

51. Is it (True, False) that deaths from accidents and natural disasters provide an effective control of population growth?

52. Make some actual comparisons between numbers of deaths due to natural disaster and replacement time through natural increase.

Disaster	Deaths	Replacement Time Through Natural Increase
a. Car accidents	________________	____________________
b. Tidal wave in India	________________	____________________
c. Vietnam War	________________	____________________

53. In summary, the population explosion is the result of:

a. increasing birth rates. (True, False)
b. decreasing death rates. (True, False)

THE SOLUTION: LOWER FERTILITY RATES

54. The only humane way to control population growth is to reduce ________________ rates to a level that matches ________________ rates.

The Demographic Transition

55. The demographic transition refers to changes in ________________ rates and ________________ rates which tend to parallel development.

56. There are (One, Two, Three, Four) major phases?

57. In the following table, indicate whether there is increase [+], decrease [-], or stability [0] in the birth rate, death rate, and population growth during each phase.

PHASE	BIRTH RATE	DEATH RATE	POPULATION
I	[ ]	[ ]	[ ]
II	[ ]	[ ]	[ ]
III	[ ]	[ ]	[ ]
IV	[ ]	[ ]	[ ]

58. Highly developed countries are near or at the end of the phase _________ where death rates are (High, Low) and birth rates are (High, Low) so that population is (High, Low, Stable).

59. Less developed countries, however, are in/at phase _________ of demographic transition where death rates are (High, Low), but birth rates are still (High, Low) and population is (High, Low, Stable).

Reasons for Disparity in Fertility Rates Between Developed and Developing Countries

60. Fertility is dependent on two main factors. List them.

a. ______________________ b. ______________________

61. Along with emotional factors, your desire regarding number of children is influenced by ________________ and ________________ factors.

62. List the social and economic factors that may influence the number of children desired.

a. ________________, b. ________________, c. ________________

d. ________________, e. ________________, f. ________________

63. Indicate whether the following list of social and economic factors would **most likely** have a positive [+], negative [-], or no [0] influence on a person or couple living in an urban area of a highly developed country (HDC), the United States, with all this country has to offer. Then fill in the table from the point of view of a poor person living on a small farm in a less developed country (LDC).

Reason	Rich person in HDC	Poor person in LDC
To care for me in periods of sickness and in old age.	[ ]	[ ]
To help with chores.	[ ]	[ ]
To contribute to the family income.	[ ]	[ ]
Because of my religious beliefs.	[ ]	[ ]
Because of the tradition in my family (pressure from family).	[ ]	[ ]
To prove myself a real man/woman.	[ ]	[ ]
Delaying having children for educational or career purposes.	[ ]	[ ]
Love and emotional fulfillment.	[ ]	[ ]

64. Is it (True, False) that a severe population explosion never developed in what are now industrialized countries?

65. Is it (True, False) that industrialization provides socioeconomic incentives to reduce fertility?

KEY WORDS AND CONCEPTS

Highly developed
less developed
crude birth rate
natural increase
survivorship
postreproductive
total fertility
population profile
growth momentum
demographic
developed
developing
crude death rate
doubling time
prereproductive
longevity
demographic transition
population structure
urbanization
demographer
replacement fertility
gross national product per capita
debt crisis

SELF TEST

Circle the correct answer to each question

1. The world population is currently (1987) about

   a. 0.5 billion.
   b. 1 billion.
   c. 2 billion.
   d. 5 billion.

2. The maximum number of people the Earth can support

   a. is about 5 billion.
   b. will depend on ecological management of the Earth's resources.
   c. will depend on the ecological regard of lifestyle.
   d. both b and c

3. Control of the world population will be best approached by

   a. reducing the birth rate
   b. improving ecological management of the Earth's resources
   c. improving the quality of life of the poor
   d. all the above

4. Highly developed nations have just 25 percent of the world's population but they hold nearly what percent of the world's economic wealth?

   a. 90, b. 80, c. 60, d. 40

5. Population changes are figured by

   a. number of deaths during a given time period minus number of births during the same time period.
   b. number of births during a given time period minus number of deaths during the same time period.
   c. number of births during a given time period minus the infant mortality rate during the same time period.
   d. the infant mortality rate during a given time period minus the number of births during the same time period.

6. Which factor does not play a part in the natural increase or decrease in population of a given country?

   a. number of deaths
   b. emigration and immigration
   c. number of births
   d. infant mortality

7. Crude birth and death rates enable one to compare

   a. the natural increase (or decrease) of populations.
   b. relative rates of immigration and emigration.
   c. the proportions of fertile women in different populations.
   d. the cause of death.

8. The most significant factor underlying the population explosion is

a. increased sexual freedom.
b. people living into their 80s and 90s as opposed to their 60s and 70s.
c. a decrease in infant and childhood mortality.
d. increased attractiveness of the opposite sex.

9. Population growth can be most effectively and acceptably controlled by

a. restricting further medical research aimed at extending longevity.
b. letting accidents and natural disasters take their course.
c. letting new diseases such as AIDS take their course.
d. reducing fertility rates.

10. People in less developed countries tend to have more children because of

a. their background and tradition of large families.
b. lack of other opportunities for self-development.
c. lack of social security plans.
d. all of the above

11. Factors that lead people in developed countries to have fewer children are

a. social security and pension plans.
b. children represent an economic liability.
c. women have many educational and career opportunities aside from child-rearing.
d. all of the above

12. In the phases of demographic transition, having less children and a low death rate is generally seen in Phase

a. I, b. II, c. III, d. IV

13. A population profile shows

a. standard of living.
b. causes of death.
c. numbers of people in each 5-year age group.
d. factors that control fertility rates.

14. The population profile of a developed country has the shape of a

a. column, b. pyramid, c. upside down pyramid, d. football on its point

15. Population growth in developing countries is approaching

a. an explosion, b. stability, c. a crash, d. a growth momentum

16. In the next 50 years, developed countries will probably represent what percent of the world's population?

a. 10, b. 20, c. 30, d. 40

17. Which of the following statements is not likely concerning countries faced with a debt crisis?

a. They pay more than half of all their earnings in interest.
b. They can obtain unlimited credit from lending agencies.
c. They are faced with high unemployment.
d. Their inflation rates top 100 percent.

18. Which of the following is the most acceptable and effective long-term control on population growth?

a. have more wars
b. improve sanitation
c. increase pre-reproductive deaths
d. decrease birth rates

19. Which of the following factors would have a negative impact on family planning by someone in a developed country but a positive impact on someone in a less developed country?

a. The need for children to help with chores.
b. The need for children to contribute to family income.
c. Pressure from the family.
d. The desire for an education rather than children.

20. What prevented a severe population explosion from occurring in industrialized countries?

a. Early successes in decreasing infant mortality.
b. Early development of mechanization and industrialization.
c. The availability of jobs in the cities.
d. All of these were contributory factors.

ANSWERS TO STUDY QUESTIONS

1. million; 2. slow, explosive; 3. a. 30, b. 100, 30, c. 30, 60, d. 15, 75, e. 12, 87, f. 12, 99; 4. explosive; 5. rapid degradation or destruction; 6. economic; 7. ecological; 8. a. highly developed, b. moderately developed, c. low income; 9. 25, 80; 10. 75, 20; 11. 10, 15; 12. unable to afford adequate food, shelter, and clothing; 13. billion; 14. c; 15. 25 to 35; 16. stability; 17. fertility; 18. -, 0, +; 19. 2; 20. profile; 21. 2, >2, <2; 22. a; 23. momentum; 24. 80; 25. 10; 26. richer, poorer; 27. living; 28. true; 29. economic growth - population growth; 30. all true; 31. all +; 32. false; 33. disease, war, family and social conditions, economic, religious traditions, moral beliefs; 34. births, deaths; 35. crude; 36. deaths, 1000; 37. natural; 38. growth; 39. lower; 40. 70; 41. short; 42. 40-80, slow; 43. die; 44. vaccinations, improved sanitation; 45. infants; 46. survivorship; 47. before, postreproductive; 48. pre-; 49. false; 50. false; 51. false; 52. (50,000, 10 days), (half million, 30 days), (45,000, 10 days); 53. true, false; 54. birth, death; 55. birth, death; 56. four; 57. (+, +, 0), (+, -, +), (-, -, 0), (-, -, 0); 58. III, low, low, stable; 59. II, low, high, high; 60. number of children desired, availability of contraceptives; 61. social, economic; 62. assets vs. liability, old-age security, educational and career opportunities, status of women, religious beliefs, availability of contraceptives; 63. (0, +), (0, +), (0, +), (0, +), (-, +), (-, 0), (+, 0), (+, +); 64. true; 65. true

---

ANSWERS TO SELF TEST

1. d; 2. d; 3. d; 4. b; 5. b; 6. b; 7. a; 8. c; 9. d; 10. d; 11. d; 12. c; 13. c; 14. a; 15. b; 16. a; 17. b; 18. d; 19. c; 20. d

CHAPTER 6

ADDRESSING THE POPULATION PROBLEM

Chapter 5 gave you a basic understanding of those factors responsible for explosive human population growth in some countries and a stable, or nearly stabilized, growth in other countries. You may recall that the fundamental resolution to population growth is to control fertility rate. However, in many countries, particularly the less developed nations, there are strong socioeconomic factors that preclude a lowered fertility rate. People in less developed countries may want to have fewer children, but they cannot afford to given the need for a larger family work force, the dependence on children to support their parents when they grow old, and social pressures to have large families. Sometimes, just plain ignorance of family planning leads to high fertility rates.

The brutal realities of the impact of explosive population growth on people's quality of life are the focus of Chapter 6. It is important for us to recognize that uncontrolled human population growth and the associated environmental degradation affects, either directly or indirectly, all people and nations of the world. There is no difference between a person living in India or the United States regarding their basic needs, i.e., food, water, housing, health, an education, and a job. People living in highly developed countries take these basic needs for granted because they are so readily available. People in less developed countries will attempt to achieve these qualities of life regardless of the environmental consequences, (e.g., cutting down the rain forests to farm or graze cattle). Chapter 6 addresses the population problem in terms of improving the lives of people, reducing fertility, and protecting the environment.

STUDY QUESTIONS

1. In order to break the cycle of *impoverished people producing an impoverished environment that in turn produces more impoverished people* we need to strive toward three world-wide conditions. List these three conditions.

   a. ______________________, b. ______________________, c. ______________________

IMPROVING LIVES OF PEOPLE

INCREASING FOOD PRODUCTION: SUCCESSES AND LIMITS

2. The most important prerequisite toward improving the human condition is to make sure that everyone has adequate ________________.

3. In 1798, Thomas Malthus predicted

   a. the value of fertilizers, b. the effects of pesticides, c. world wide famine
   d. higher fertility rates in less developed countries

4. Is it (True, False) that agricultural production has kept pace with and even exceeded population growth?

5. Between 1950 and 1984, world grain production moved ahead of population growth by ________ percent.

6. Indicate whether the following factors have increased [+] or decreased [-] agricultural production in the last 40 years.

   [ ] Bringing additional land into cultivation.
   [ ] Increasing use of fertilizer.
   [ ] Increasing use of irrigation.
   [ ] Increasing use of chemical fertilizers.
   [ ] Substituting new genetic varieties of grains.

<u>Land Area Devoted to Grain Production</u>

7. Much of the new grain land brought into production from 1950 to 1981 has now been abandoned because of

   a. erosion. (True, False)
   b. water depletion. (True, False)
   c. urban development. (True, False)
   d. highway development. (True, False)

8. The total area in grain production fell about ________ percent from 1981 to 1988.

<u>Irrigation</u>

9. Even though irrigated acreage increased about 2.6 times from 1950 to 1980, irrigation is increasing at a slower pace due to

   a. lack of new water sources. (True, False)
   b. depletion of existing ground water resources. (True, False)
   c. waterlogging and accumulation of salts in the soil. (True, False)

<u>Fertilizer</u>

10. From 1950 to 1984, fertilizer use grew about ________ fold.

11. Is it (True, False) that there is an optimum amount of fertilizer required for maximum growth?

12. Adding more fertilizer than the optimum required leads to

   a. waste. (True, False)
   b. increased plant vulnerability to pests. (True, False)
   c. pollution. (True, False)

<u>Pesticides</u>

13. The use of pesticides after World War II

   a. provided better pest control. (True, False)
   b. increased crop yields. (True, False)
   c. led to pest resistance. (True, False)
   d. caused adverse side effects to human and environmental health. (True, False)

High Yielding Varieties

14. The new varieties of wheat and rice produced in the 1960s required more ______________ and ______________ but gave yields ______________ to ______________ when compared to traditional varieties.

15. The high world wide production of wheat and rice resulting from the new genetic varieties was hailed as the ______________ revolution.

Climate

16. Three severe droughts in 19 ______, 19 ______, and 19 ______ drastically affected harvests.

17. Is it (True, False) that droughts are predicted to become increasingly commonplace?

18. Is it (True, False) that per capita productions of grains and fish are heading downward?

THE HUNGRY: A PROBLEM OF FOOD-BUYING CAPACITY

19. Simply weighing current world human population against world agricultural capacity, is it (True, False) that the world cannot produce enough food and, therefore, we are on the brink of a world famine?

20. Is it (True, False) that the world produces food surpluses?

21. Is it (True, False) that all people of the world have equal food-buying power?

22. Indicate whether the following economic factors have a [+] or [-] effect on peoples' ability to pay for food.

    [ ] Growing cash crops.
    [ ] Eating high on the food chain.
    [ ] Putting land back in the hands of peasant farmers.

23. Eating a high meat diet

    a. is necessary for good health. (True, False)
    b. consumes less grain than would be consumed if grains were eaten directly. (True, False)
    c. consumes about the same amount of grain as if grains were eaten directly. (True, False)
    d. consumes 5 to 10 times more grain than if grains were eaten directly. (True, False)

24. If people in developed countries switched from high meat diets to eating mostly grain and vegetables

    a. there would be enough grain to feed another 2 to 3 billion people. (True, False)
    b. there would be a great increase in ill health and mortality from cancer and heart disease. (True, False)
    c. there may be improved health and a decrease in mortality from cancer and heart disease. (True, False)

25. Is it (True, False) that many developing countries, where malnutrition is present, devote a large portion of their agricultural capacity to cash crops?

FOOD AID AND THE UTTERLY DISMAL THEOREM

26. Indicate whether the following statements are true [+] or false [-] concerning the overall effects of providing food aid to countries in need.

    [ ] Providing food actually alleviates chronic hunger in developing countries.
    [ ] It undercuts prices for domestically produced food sold in local markets.
    [ ] The entire local economy deteriorates.
    [ ] It contributes to environmental and ecological deterioration.

27. The continuing cycle of economic disruption resulting from food aid to developing countries is called the ________________ ________________ theorem.

28. The reason that food aid does not alleviate the problem of chronic hunger in less developed countries is because it (Disrupts, Stimulates) the local economy and leads to a/an (Increase, Decrease) in local food production.

29. Food aid tends to (Improve, Aggravate) the problem of chronic hunger.

30. To alleviate poverty and hunger, it is necessary to focus on ________________ development and on reducing ________________ rates.

ECONOMIC DEVELOPMENT

31. Promoting economic development in developing countries

    a. is a humanitarian goal that produces its own rewards. (True, False)
    b. reduces fertility rates. (True, False)
    c. promotes prosperity. (True, False)

32. List the two kinds of economic development projects.

    a. ____

    b. ____

Large-scale Centralized Projects

33. Give three examples of large-scale centralized projects.

    a. ________________, b. ________________, c. ________________

34. Indicate whether the following statements are true [+] or false [-] concerning large-scale centralized projects.

    [ ] Relatively easy to administer, measure, monitor, and to demonstrate end products.
    [ ] Result in substantial increases in the gross national product.
    [ ] Result in substantial increases in wealth for all residents of the developing country.
    [ ] Centralized projects actually aggravate poverty.
    [ ] Requires the use of modern machinery and technology.
    [ ] Increases diversity in crops and methods of farming.
    [ ] Promotes nonsustainable agriculture.
    [ ] Centralized projects increase the debt crisis in developing countries.

35. Major funding organizations are now considering loan requests from developing countries in terms of ________________ development.

Decentralized Projects--Appropriate Technology

36. Appropriate technology projects should

a. not upset the existing social structure. (True, False)
b. involve a high degree of training or skill. (True, False)
c. utilize local resources. (True, False)
d. utilize centralized workplaces. (True, False)
e. allow individuals to develop self-reliance. (True, False)

37. The key to appropriate technology is a form of aid that enables people to be more productive within the framework of

a. their existing social and economic system. (True, False)
b. a new governmental system. (True, False)
c. a new socioeconomic system. (True, False)

38. Indicate whether the following are examples of large-scale centralized [LC] or appropriate technology [AT].

[ ] Textile mill
[ ] Handloom
[ ] Large apartment complex
[ ] Handmade brick dwelling
[ ] More efficient wood stoves

Appropriate Technology and Agriculture

39. Appropriate technology can be applied to agriculture by

a. use of waste as mulches. (True, False)
b. use of chemical fertilizers. (True, False)
c. introduction of poly-cropping. (True, False)
d. planting large acreages of monoculture. (True, False)
e. producing methane gas using anaerobic digesters. (True, False)
f. using more gasoline and fossil fuels. (True, False)
g. feeding fish agricultural wastes. (True, False)
h. reforestation. (True, False)

40. Is it (True, False) that people in less developed countries need power plants?

REDUCING FERTILITY

41. Fertility rates may be reduced substantially through

a. ________________________________

b. ________________________________

c. ________________________________

42. There is a (Strong, Weak) correlation between per capita income and total fertility rate.

43. Is it (True, False) that leaders of less developed nations support reducing fertility?

44. Is it (True, False) that people of less developed countries want fewer children?

LEADERS OF LESS DEVELOPED NATIONS SUPPORT REDUCING FERTILITY

45. More than __________ nations representing over _________ percent of the world's population now have family planning programs.

46. __________ percent of the funding for family planning programs is provided by (Highly, Less) developed countries.

PEOPLE OF LESS DEVELOPED COUNTRIES WANT FEWER CHILDREN

47. An extensive survey of women in less developed countries revealed that

a. women, for various reasons, still want as many children as they can bear. (True, False)
b. most women want fewer children but are not using effective contraceptives. (True, False)
c. there is much room for expanding family planning services. (True, False)

48. An extensive survey of women in less developed countries revealed that ________ percent wanted no further children, and an additional __________ percent wanted to delay their next pregnancy for at least two years.

49. Of the women who wanted to delay the next pregnancy, _________ did **not** know of any family planning source.

50. Is it (True, False) that lack of reading and writing skills will preclude effective use of family planning resources?

EFFECTIVENESS OF FAMILY PLANNING, HEALTH CARE, AND EDUCATION

51. Indicate whether the following practices are [+] or are not [-] characteristic of family planning services and information.

[ ] Counseling on reproduction and contraceptive techniques.
[ ] Supplying contraceptives.
[ ] Counseling on pre- and post-natal health of mother and child.
[ ] Counseling on the health advantages of spacing children.
[ ] Promoting as many abortions as possible.
[ ] Nursing a baby for as long as possible.
[ ] How to prevent unwanted pregnancies.

52. Health care programs that are well received and effective

a. promote better maternal health. (True, False)
b. promote fewer but healthier children. (True, False)

53. One of the biggest handicaps to family planning is low levels of _________________ among women of less developed countries.

54. Literacy of the mother is particularly important because she must be able to read in order to follow printed directions regarding _________________, ___________________, and ____________________.

55. Providing educational opportunities for children will also tend to reduce birth rates because

a. parents see education as a way for their children to achieve better lives. (True, False)
b. parents tend to desire fewer children in order to allocate more of their resources to educating a few. (True, False)
c. education results in significant reductions in fertility at a very modest cost. (True, False)

ADDITIONAL ECONOMIC INCENTIVES

56. Indicate whether the following would be incentives [I] or deterrents [D] toward increasing birth rates.

[ ] Taxation, [ ] Education, [ ] Housing, [ ] Maternity Care

57. A country that has implemented the most extensive set of economic incentives and deterrents regarding fertility is ________________.

58. By using economic incentives and deterrents, China had reduced its fertility rate from _________ in the mid-1970s to _________ in 1982 and to ________ currently.

59. A government can reduce birth rates, without unduly alienating its citizens, by

a. mandating sterilization after a person has had one or two children. (True, False)
b. providing economic incentives toward delaying and/or limiting childbearing. (True, False)
c. providing economic deterrents toward having more than a certain number of children. (True, False)

PROMOTING FAMILY PLANNING AND THE ABORTION CONTROVERSY

60. Family planning has become the center of extreme controversy between __________________ and __________________ groups.

61. ___________________ groups advocate the position that women ought to have the right to choose whether or not to have a child including the option to have a safe abortion.

62. ___________________ groups exclude the option of abortions regardless of the circumstances.

63. Which of the following would **not** be a result of legal abortions?

a. Needless deaths of thousands of women resulting from illegal or self-induced abortions.
b. Protection of women from pregnancies that may impair their future health.
c. A realization of the full scope of family planning services.

CONTRACEPTIVE TECHNOLOGY

64. A perfect contraceptive

a. does not require remembering. (True, False)
b. is fully and immediately reversible. (True, False)
c. has no painful or harmful side effects. (True, False)
d. is freely available to all women. (True, False)
e. is effective 100% of the time. (True, False)

65. Indicate whether the following technologies are contraceptive [C] or abortive [A].

[ ] Norplant [ ] RU486

COSTS OF FAMILY PLANNING

66. There are over __________ million people in the developing world who want to limit the size of their families but do not have the knowledge or means to do so.

67. The estimated costs for providing family planning services world wide are $_________ billion in 1989 rising to $_________ billion in 1998 with the U.S. share being $_________ million per year at the start.

68. In contrast, world wide expenditures for armaments are more than a _______________ dollars per year.

69. Long-term security of a nation depends as much on

a. stabilizing population
b. protecting the environment
c. both a and b

as on protection from outside enemies.

70. Is it (True, False) that the present U.S. policy on providing financial support for family planning includes provision of funds for needed abortions?

KEY WORDS AND CONCEPTS

starvation	malnutrition	food production capacity
food buying capacity	cash crops	eating high on the food chain
Utterly dismal theorem	debt crisis	large-scale centralized projects
appropriate technology	per capita income	decentralized projects
economic incentives	family planning service	economic deterrents
contraceptives	abortions	Norplant
RU486	fertility	

SELF-TEST

Circle the correct answer to each question.

1. Addressing the population problem will require

a. improving the lives of people.
b. reducing fertility.
c. protecting the environment.
d. all of the above are required

2. The most important prerequisite toward improving the human condition is to make sure that everyone has adequate

a. education, b. numbers of children, c. nutrition, d. jobs

3. In 1798, Thomas Malthus predicted

a. the increasing value of fertilizer.
b. the effects of pesticides on the natural ecosystem.
c. world wide famine.
d. higher fertility rates in less developed countries.

4. Which of the following was not a factor leading to increased agricultural production in the last 40 years?

a. Bringing additional land into cultivation.
b. People in highly developed countries eating lower on the food chain.
c. Increased use of inorganic fertilizers.
d. Substituting new genetic varieties of grains.

5. Much of the new grain land brought into production from 1950 to 1981 has now been abandoned because of

a. erosion
b. water depletion
c. urban and highway development
d. all of the above

6. Which of the following is not a reason why irrigation is increasing at a slower pace?

a. Inability to find new water sources.
b. Depletion of existing ground water sources.
c. A realization by growers that they need to conserve ground table water.
d. Waterlogging and accumulation of salts in the soil.

7. Adding more fertilizer than the optimum leads to

a. waste, b. increased plant vulnerability to pests, c. pollution, d. all of these

8. Which of the following was not a result of pesticide use after World War II?

a. Prior testing of potential adverse environmental effects.
b. Increased crop yields.
c. Pest control.
d. Pest resistance.

9. The green revolution did not provide

a. new varieties of wheat and rice.
b. enormous increases in food production in less developed countries.
c. increased ability for poor people to buy food.
d. advanced technologies in food production.

10. Which of the following would have the greatest positive impact on peoples' ability to buy food?

a. Putting land into the hands of peasant farmers.
b. Growing cash crops.
c. Eating higher on the food chain.
d. Development of large-scale centralized farming technology.

11. Eating a high meat diet

a. is necessary for good health.
b. consumes less grain than would be consumed if grains were eaten directly.
c. consumes about the same amount of grain as if grains were eaten directly.
d. consumes 5 to 10 times more grain than if grains were eaten directly.

12. Which of the following would not be a result of people in developed countries switching from high meat diets to eating mostly grain and vegetables.

a. There would be enough grain to feed another 2 to 3 billion people.
b. They would experience better health and a longer life.
c. They would require protein supplements to replenish lost meat protein.
d. Less developed countries would commit less of their agricultural capacity to cash crops.

13. The "Utterly Dismal Theorem" refers to

a. the theory that collapse of the biosphere is inevitable.
b. the theory that wide spread famine is inevitable.
c. observations that supplying free food worsens the problem it is intended to alleviate.
d. anticipated loss of habitat from overpopulation.

14. It has been observed that giving free food to a country with a chronic hunger problem:

a. aggravates the hunger problem.
b. leads to an increase in local food production.
c. stabilizes the local economy.
d. all of the above

15. The most effective and acceptable way to solve the world hunger problem would be to

a. provide free food.
b. provide economic development to combat poverty.
c. let the problem run its course through starvation.
d. allow abortions and euthanasia.

16. Which of the following is not a result of large-scale centralized projects?

a. They are relatively easy to administer and to demonstrate end products.
b. They result in substantial increases in the gross national product.
c. They result in substantial increases in the wealth of all residents.
d. They require the use of modern machinery and technology.

17. The utilization of local resources without upsetting the social structure is characteristic of

a. centralized projects, b. appropriate technology,
c. government projects, d. modern agricultural technology

18. Declines in birth rates have been found to correlate most closely with

a. increase in average per capita income.
b. increase in the level of education achieved by women.
c. having and raising one child.
d. availability of contraceptives.

19. Family planning services do not provide

a. counseling on reproduction and contraceptive techniques.
b. contraceptives.
c. unlimited abortions.
d. counseling on the health advantages of spacing children.

20. Which of the following is a "nearly perfect" contraceptive?

a. I.U.D., b. Norplant, c. RU486, d. birth control pill

ANSWERS TO STUDY QUESTIONS

1. improve lives of people, reduce fertility, protect the environment; 2. nutrition; 3. c; 4. true; 5. 40; 6. all +; 7. all true; 8. 7; 9. all true; 10. 9; 11. true; 12. all true; 13. all true; 14. fertilizer, water, double, triple; 15. green; 16. 80, 83, 88; 17. true; 18. true; 19. false; 20. true; 21. false; 22. -, -, +; 23. false, false, false, true; 24. true, false, true; 25. true; 26. -, +, +, +; 27. utterly dismal; 28. disrupts, decrease; 29. aggravate; 30. economic, fertility; 31. all true; 32. large-scale centralized projects, decentralized projects; 33. hydroelectric dams, industrial plants, high-speed highways; 34. +, +, -, +, +, -, +, +; 35. sustainable; 36. true, false, true, false, true; 37. true, false, false; 38. LC, AT, LC, AT, AT; 39. true, false, true, false, true, false, true, true; 40. true; 41. family planning, health care, education; 42. strong; 43. true; 44. true; 45. 100, 95; 46. 75, less; 47. false, true, true; 48. 50, 25; 49. 60; 50. true; 51. +, +, +, +, -, +, +; 52. true, true; 53. education; 54. health care, nutrition, contraceptives; 55. all true; 56. all D's; 57. China; 58. 4.5, 2.6, 2.4; 59. false, true, true; 60. right-to-life, pro-choice; 62. pro-choice; 63. c; 64. all true; 65. C, A; 66. 400; 67. 2.4, 4.1, 600; 68. trillion; 69. c; 70. false

---

ANSWERS TO SELF TEST

1. d; 2. c; 3. c; 4. b; 5. d; 6. c; 7. d; 8. a; 9. c; 10. a; 11. d; 12. c; 13. c; 14. a; 15. b; 16. c; 17. b; 18. b; 19. c; 20. b

CHAPTER 7

SOIL AND THE SOIL ECOSYSTEM

If one were to name the one most important endeavor underlying the support of civilization it would have to be a stable agricultural production. No society can maintain itself for long in a civilized state without an ample reliable food supply for its citizens and such a supply can only come from the artificial propagation of various plants and animals, i.e. agriculture. Propagation of plant species is even more fundamental than that of animals because plants are always at the bottom of the food chain.

In turn, the two resources that are most fundamental in supporting plant production are fertile soil and adequate water. We take solar energy as a given since it remains essentially constant and humans can do little to manage it in one way or another. Both fertile soil and water supplies are what may be termed renewable resources since with proper management, they can be maintained, used and reused indefinitely. But, without proper management, they can be depleted and/or destroyed. Indeed, there is much evidence to suggest the many ancient extinct civilizations came to their downfall as a result of depleting their soil and/or water resources.

Given such bitter lessons of the past, it would seem that we would have arrived at the point of appreciating the fundamental importance of soil and water resources, and that we would be maintaining them appropriately especially in the face our increasing population discussed in Part II. Unfortunately, such is not the case. While we have the understanding and the technology to maintain soil and water resources, in too many cases we are not doing so. We are exploiting and squandering these resources in a manner that produces short-term gain but in the long-term leaves them depleted and destroyed.

A sustainable society can only be one which satisfies its needs without diminishing the prospects of coming generations. (Hussain Muhammad Ershad, President of Bangladesh, 1987) Clearly, the way we are depleting and destroying water and soil resources does not hold well for future generations. Current society is not sustainable if we persist with current attitudes and methods toward depletion and destruction of these precious resources. If we wish to sustain our society we must devote considerably more thought and attention to the maintenance of soil and water resources.

Fortunately, we do have the necessary knowledge to manage and maintain soil and water resources. We only need to apply this knowledge through appropriate actions.

STUDY QUESTIONS

PLANTS AND SOIL

CRITICAL FACTORS OF THE SOIL ENVIRONMENT

1. The soil environment must supply the plant or at least its roots with ________________, ____________________, and ____________________.

Mineral Nutrients and Nutrient-Holding Capacity

2. Four important mineral nutrients required by the plant are ________________, ________________, ________________, and ________________.

3. Three non-nutrient inputs into the soil are ________________, ________________, and ________________.

4. The original source of mineral nutrients is from the breakdown of ________________ through the process of ________________.

5. Nutrient ions may be washed away by water percolating through the soil, a process called ________________.

6. The capacity of the soil to bind and hold the nutrient ions is called the soil's ________________ capacity.

7. Is it (True, False) that the process of weathering is fast enough or that sources are great enough to provide the nutrient needs to sustain vigorous plant growth without other sources?

8. In natural ecosystems, the greatest source of mineral nutrients for sustaining plant growth is from ________________ of nutrients.

9. In agricultural situations there is an unavoidable removal of nutrients through the ________________ of crops. Such nutrients may be replaced through additions of ________________ fertilizers.

10. ________________ fertilizers consist of plant and animal wastes.

Water and Water-Holding Capacity

11. A plant loses a large amount of water by evaporation through its leaves through a process called ________________.

12. If the water lost in transpiration is not replaced the plant will ________________.

13. Is it (True, False) that plants need a more or less continuous supply of water to replace that lost in transpiration?

14. Since it rains only intermittently, the plants depend on a reservoir of water held in the ________________ soil.

15. In addition to the frequency and amount of precipitation, the amount of water that is actually available to plants will depend on

    a. the amount of water that infiltrates versus running off. (True, False)
    b. the amount of water that is held in the soil versus percolating through. (True, False)
    c. the amount of water that evaporates from the soil surface. (True, False)

16. Indicate whether the following conditions need to be maximized [+] or minimized [-] in order to provide plants with the largest amount of available water.

    [ ] runoff, [ ] infiltration, [ ] water holding capacity, [ ] surface evaporation

Oxygen and Aeration

17. The ability of soil to allow the diffusion of oxygen into the soil and carbon dioxide out of the soil is referred to as soil _______________.

18. The roots of most plants have access to oxygen by

a. diffusing it through the soil.
b. transfer through plant leaves.
c. having special breathing roots at the soil surface.

19. Packing down the soil until air space becomes too limited to allow diffusion is called _______________.

20. Saturating soil spaces with water is called _______________.

Relative Acidity (pH)

21. pH is a measure of relative _______________ and _______________.

22. Most plants require a soil environment which is

a. acidic, b. basic or alkaline c. close to neutral

23. Neutral is expressed by a pH of ________.

Salt and Osmotic Pressure

24. The relative salt concentration inside and outside the cell membrane is determined by a feature called _______________ balance.

25. The movement of water across a cell membrane toward a higher salt concentration is a phenomenon called _______________.

26. If the salt concentration outside plant cells is higher than inside the cells, the plant cell (Can, Cannot) absorb water.

27. Most plants require (Fresh, Salty) water.

GROWING PLANTS WITHOUT SOIL

28. The culturing of plants without soil is called _______________.

29. Plants cultured hydroponically

a. are just as nutritious as those grown on good soil. (True, False)
b. cost less to produce than plants grown on good soil. (True, False)
c. demonstrate that preserving the soil ecosystem is not necessary. (True, False)

THE SOIL ECOSYSTEM

30. List the three ingredients required of productive topsoil.

a. _______________, b. _______________, c. _______________

Soil Texture--Size of Mineral Particles

31. The size of soil particles is defined as soil _______________.

32. Rank (1 = largest, 2 = middle, 3 = smallest) the below examples of the three main size categories of soil mineral particles.

[ ] clay, [ ] silt, [ ] sand

33. **Loam** has a mineral composition of roughly _______ percent sand, ________ percent silt, and ________ percent clay.

34. As a soil becomes more coarse (average particle size increasing from clay to coarse sand), indicate whether each of the following soil properties will increase [+], decrease [-], or remain the same [0].

[ ] Infiltration
[ ] Surface runoff
[ ] Water-holding capacity
[ ] Aeration
[ ] Nutrient-holding capacity

35. The workability of clay soils will be (More, Less) difficult when compared to sandy soils.

Detritus, Soil Organisms, Humus, and Topsoil

36. Distinguish between **humus** [H] and **detritus** [D] in the following definitions.

[ ] Accumulation of dead leaves and roots on and in the soil.
[ ] The residue of undigested organic matter that remains after dead leaves and roots have been consumed.

37. The base (1st trophic level) of the food web in soil ecosystems is

a. producers, b. detritus, c. humus, d. decomposers

38. The second trophic level of the food web in soil ecosystems is

a. producers, b. detritus, c. humus, d. decomposers

39. As soil organisms feed on the detritus and reduce it to humus, their activity also mixes and integrates humus with mineral particles developing what is called soil _______________.

40. The feeding and burrowing activities of organisms, which are feeding directly and indirectly on detritus

a. break the detritus down to humus. (True, False)
b. mix and integrate the humus with the mineral particles of the soil. (True, False)
c. cause the soil to become loose and clumpy. (True, False)
d. result in the formation of topsoil. (True, False)

41. In comparison to underlying subsoil, topsoil

a. is generally darker in color. (True, False)
b. has a loose, clumpy structure whereas subsoil is more compacted. (True, False)

42. Indicate whether the presence of humus and its clumpy structure in topsoil does [+] or does not [-] improve the following soil characteristics.

[ ] infiltration
[ ] water-holding capacity
[ ] nutrient-holding capacity
[ ] markedly superior aeration
[ ] workability

Other Soil Biota

43. A healthy soil ecosystem contains

a. **mycorrhizae** fungi which have a beneficial symbiotic relationship with roots by helping them absorb nutrient. (True, False)
b. certain organisms which attack root parasites. (True, False)

Organic vs Inorganic Fertilizer

44. Is it (True, False) that there is **no** evidence that organic compounds from the soil are required for the plant's nutrition?

45. Inorganic (chemical) fertilizers

a. can be added to the soil efficiently and economically. (True, False)
b. substitute for additions of detritus. (True, False)
c. causes soil organisms to starve thus depleting humus content. (True, False)
d. causes soil structure to collapse. (True, False)

46. What agency provides soil testing services and recommendations regarding needed nutrients?

____________________ ____________________ ____________________

MUTUAL INTERDEPENDENCE OF PLANTS AND SOIL

47. Is it (True, False) that green plants support the soil organisms by being the direct or indirect source of all their food?

48. Is it (True, False) that soil organisms support green plants by making the soil more suitable for their growth?

49. As humus and the clumpy soil structure break down, indicate whether the following increase [+], decrease [-], or remain the same [0].

[ ] infiltration
[ ] aeration
[ ] runoff
[ ] nutrient-holding capacity
[ ] water-holding capacity
[ ] leaching of nutrients
[ ] the ability of the soil to support plants

LOSING GROUND

BARE SOIL, EROSION, AND DESERTIFICATION

50. The loss of soil is called ________________.

51. Erosion refers to soil particles being picked up by ____________________ or ____________________.

52. Name the different forms of erosion shown in figures 7-21, 7-22, and 7-23.

    a. ______________________________ b. ___________________________

    c. ______________________________ d. ___________________________

53. For each of the following categories of soil particles, indicate the order [1 = first, 5 = last] that they would be removed through wind or water erosion.

    [ ] clay, [ ] humus, [ ] silt, [ ] fine sand, [ ] coarse sand and stones

54. The conversion of productive land into a desert is called __________________.

CAUSES OF LOSING GROUND

55. List the three major cause of soil erosion and desertification.

    a. __________________, b. ___________________, c. ___________________

Overcultivation

56. The traditional agricultural practice that has resulted and still is resulting in extensive erosion is __________________.

57. The basic reason for plowing and cultivation is for ____________________ control.

58. The basic drawback to plowing is that soil is exposed to _________________ and _________________ erosion.

59. Is it (True, False) that plowing improves aeration and infiltration?

Overgrazing

60. When grass is grazed faster than it can regenerate itself, this process is called _________________.

61. Overgrazing leads to

    a. less detritus to generate soil humus. (True, False)
    b. a gradual mineralization of the soil. (True, False)
    c. decreased grass production. (True, False)
    d. increased soil erosion. (True, False)
    e. decreased rainfall in the region. (True, False)

Deforestation

62. A forest cover

a. breaks the fall of raindrops reducing splash erosion. (True, False)
b. allows water to infiltrate a litter-covered, loose topsoil. (True, False)
c. allows a 50 percent less runoff when compared to grasslands. (True, False)
d. reduces 45-fold the leaching of soil nitrogen. (True, False)

63. Which of the following is **not** a usual reason for clearing forests?

a. To experiment with secondary succession.
b. To permit agriculture.
c. To obtain structural wood.
d. To obtain firewood.

64. Indicate whether the following uses of wood would be predominantly characteristic of [a] developed countries, [b] developing countries, or [c] both.

[ ] To permit agriculture
[ ] To obtain structural wood
[ ] To obtain firewood

65. In some developing countries, forests are cut _________ times faster than they are able to grow back.

Irrigation, Salinization and Desertification

66. Adding water by artificial means is called ________________.

67. Irrigated water contains at least 200-500 parts per million salt which when left behind through evaporation leads to ________________.

68. _________ percent of all irrigated land has already been salinized and an additional _________ to _________ million acres are salinized each year.

DIMENSIONS OF THE PROBLEM

69. In the United States, about ________ percent of our original agricultural land has been rendered nonproductive by __________________.

70. Soils can sustain an erosion rate of up to _________ tons per acre and still remain in balance.

71. However, erosion is 2 to _________ times the tolerable rate.

72. U.S. farmers are sacrificing about _______ tons of topsoil for every ton of grain produced.

73. The most recent example of a country becoming totally desertified is __________________.

74. Loss of topsoil leads to

a. increased sediment deposits in rivers. (True, False)
b. flooding in lowlands. (True, False)
c. a continual cycle of erosion. (True, False)

PREVENTING EROSION AND DESERTIFICATION

Preventing Erosion: Traditional Techniques

75. List the four traditional methods of controlling soil erosion.

    a. ______________________, b. ______________________

    c. ______________________, d. ______________________

76. Indicate whether the following descriptions of methods of controlling soil erosion apply to [a] contour farming, [b] strip cropping, [c] shelter belts, or [d] terracing.

    [ ] Planting rows of trees around fields.
    [ ] Plowing and cultivating at right angles to the slope.
    [ ] Planting alternative strips of grass between strips of corn.
    [ ] Grading slopes into a series of steps.

Preventing Erosion: New Techniques

77. The advantages of **no-till** agriculture include

    a. erosion is minimized. (True, False)
    b. detritus remains. (True, False)
    c. saves time and energy. (True, False)
    d. minimizes soil compaction. (True, False)
    e. earlier plantings and harvests. (True, False)
    f. use of chemicals. (True, False)

78. Plants that grow, set seed, and die in a single season are called (Annuals, Perennial).

79. Plants that grow from the same root stock year after year are called (Annuals, Perennial).

80. Major crops that support humans are (Annuals, Perennial).

81. Major crops that should support humans should be switched to (Annuals, Perennial).

Limit Grazing

82. Is it (True, False) that grazing must be limited to what the range land can sustain?

Reforestation and Rehabilitation of Desertified Lands

83. A country that has been successful in bringing deserts back into agriculture is (United States, Ethiopia, Israel).

Countering Salinization

84. Is it (True, False) that drip irrigation systems will mitigate salinization?

IMPLEMENTING SOLUTIONS

85. The ravaging effects of the soil erosion during the Dust Bowl era in America led to the development of the ______________________________ in 1935.

86. The most recent soil conservation legislation is the ______________________________ of 1985.

87. The two major goals of the Food Security Act of 1985 are:

a. ____

b. ____

KEY WORDS AND CONCEPTS

mineral nutrients
nutrient-holding capacity
parent material
weathering
percolation of water
leaching
inorganic (chemical) fertilizer
organic fertilizer
compaction (of soil)
ph
osmosis
soil texture
sand
silt
clay
loam
workability
humus
soil structure
topsoil
subsoil
soil organisms
soil biota
soil ecosystem
mineralization
erosion
water-holding capacity
infiltration
runoff
transpiration
wilting
field capacity
capillary water
aeration (of soil)
rivulet erosion
gully erosion
wind erosion
water erosion
strip cropping
contour farming
terracing
shelter belts
no-till agriculture
overgrazing
desertification
clearcutting
selective cutting
irrigation
salinization
hydroponics
splash erosion
sheet erosion

SELF-TEST

Circle the correct answer to each question.

1. Plants require _________ from the soil.

a. mineral nutrients
b. water
c. air (oxygen)
d. all of the above

2. The gradual breaking down of rock into soil particles by natural factors such as freezing and thawing is called

   a. leaching, b. abrading, c. weathering, d. erosion

3. Most land plants require substantial quantities of water to replace water lost by way of

   a. transpiration, b. respiration, c. photosynthesis, d. aeration

4. Water soaking into soil is known as

   a. aeration, b. infiltration, c. capillary action, d. leaching

5. Packing down the soil very tightly around the roots of plants is most likely to

   a. aid their growth because it makes nutrients more available.
   b. aid their growth because it will make water more available.
   c. retard their growth because it will reduce aeration.
   d. retard their growth because it will reduce sunlight available to soil organisms.

6. Soil aeration, nutrient-holding capacity, and water-holding capacity are all affected by

   a. soil texture, b. presence of humus, c. soil biota, d. all of these

7. In measuring soil texture, one would determine the

   a. proportion of sand, silt, and clay-sized particles.
   b. nutrient content of the soil.
   c. water-holding capacity of the soil.
   d. infiltration rate.

8. The smallest sized particles in soil are

   a. sand, b. silt, c. clay, d. stones

9. A soil made up of a mixture of roughly 40% sand, 40% silt and 20% clay would technically be defined as

   a. topsoil, b. loam, c. subsoil, d. humus

10. As soil particle size increases, which of the following soil properties decrease?

    a. infiltration, b. surface runoff, c. water-holding capacity, d. aeration

11. Humus is

    a. one of the inorganic, mineral constituents of the soil.
    b. the collection of all the soil microorganisms.
    c. finely divided bits of leaves, twigs, and other litter.
    d. a residue of organic matter that remains after decomposition.

12. The base of the food chain in soil ecosystems is

    a. producers, b. detritus, c. humus, d. decomposers

13. Humus added to a loam soil will greatly improve

a. infiltration, b. water-holding capacity, c. nutrient-holding capacity, d. all of these

14. As humus breaks down, which of the following soil properties increase?

a. infiltration, b. aeration, c. runoff, d. nutrient-holding capacity

15. The process of soil being washed or blown away is called

a. erosion, b. leaching, c. water logging, d. compaction

16. A vegetative cover will protect soil from

a. splash erosion, b. sheet erosion, c. wind erosion, d. all of these

17. A soil conservation practice that involves a method of cultivation is

a. strip cropping, b. contour farming, c. terracing, d. shelter belts

18. Desertification is a process of

a. semi-arid grasslands becoming deserts.
b. soil becoming more and more salty.
c. forests being clearcut.
d. wind erosion.

19. Which of the following is not a cause of desertification?

a. overgrazing, b. clear cutting forests, c. reducing vegetative cover, d. all are causes

20. The disadvantage of no-till agriculture is related to

a. erosion, b. time and energy, c. chemicals, d. compaction

ANSWERS TO STUDY QUESTIONS

1. mineral nutrients, water, oxygen; 2. nitrate, phosphate, potassium, calcium; 3. silica, aluminum, oxygen; 4. rocks, weathering; 5. leaching; 6. nutrient-holding; 7. false; 8. recycling; 9. harvest, inorganic; 10. organic; 11. transpiration; 12. wilt; 13. true; 14. surface; 15. all true; 16. -, +, +, -; 17. aeration; 18. a; 19. compaction; 20. water logging; 21. acidity, alkalinity; 22. c; 23. 7; 24. water; 25. osmosis; 26. cannot; 27. fresh; 28. hydroponics; 29. true, false, false; 30. mineral particles, detritus, living organisms; 31. texture; 32. 3, 2, 1; 33. 40, 40, 20; 34. +, +, -, +, -; 35. more; 36. D, H; 37. b; 38. d; 39. structure; 40. all true; 41. all true; 42. all +; 43. all true; 44. true; 45. all true; 46. Agricultural Extension Service; 47. true; 48. true; 49. -, -, +, -, -, +, -; 50. erosion; 51. wind, water; 52. splash, sheet, rivulet, gully; 53. 4, 1, 2, 3, 5; 54. desertification; 55. overcultivation, overgrazing, deforestation; 56. plowing; 57. weed; 58. wind, water; 59. false; 60. overgrazing; 61. (a-d = true), false; 62. all true; 63. a; 64. C, C, B; 65. 5; 66. irrigation; 67. salinization; 68. 30, 2.5, 3.7; 69. 20; 70. 5; 71. 10; 72. 6; 73. Ethiopia; 74. all true; 75. contour plowing, strip cropping, shelter belts, terracing; 76. c, a, b, d; 77. (a-e = true), false; 78. annuals; 79. perennial; 80. annuals; 81. perennial; 82. true; 83. Israel; 84. true; 85. Soil Conservation Service; 86. Food Security Act; 87. convert erodible cropland into conservation reserve, implement soil conservation plans

ANSWERS TO SELF TEST

1. d; 2. c; 3. a; 4. b; 5. c; 6. d; 7. a; 8. c; 9. b; 10. c; 11. d; 12. b; 13. d; 14. c; 15. a; 16. d; 17. b; 18. a; 19. d; 20. c

CHAPTER 8

WATER, THE WATER CYCLE, AND WATER MANAGEMENT

Just outside Phoenix, which gets less than 8 inches of rain a year, the fountains shoot high into the arid air, gurgling defiance of the laws of evaporation, advertising the subdivisions whose green lawns eat further and further into the Arizona desert. But in parts of the Texas High Plains, the water that made farms green with cotton and melons is giving out; land is now brown with tumbleweeds, which farsighted agronomists are investigating as a cash crop. And in stately Greenwich, Conn., surely one of the last places from which civilization will vanish, suburban matrons guard water like Bedouins, and town officials lay plans for slit-trench latrines against the not-too-distant day when the reservoirs may run dry. (Newsweek, Are We Running Out of Water? February 23, 1981, page 26)

No life can exist without water. Water is fundamental in the support of all ecosystems. Water is fundamental in the production of all food crops and animals. We use water to flush away wastes. In cooling we often use water to carry away waste heat. We also require water for drinking, recreation, and its aesthetic charm in fountains and falls. It would be hard to argue that any resource is more precious than water. Yet, we are squandering and depleting our water resources. We are making water unusable and we are making it unable to support desired ecosystems by our pollution.

There is little question that our apathy toward, our lack of understanding of, and our blatant disregard for water resources is placing all life on Earth in jeopardy. We desperately need to understand the nature of water and our water resources. More importantly, we need to use this understanding with care and wisdom to conserve and care for our water resources. Our lives will depend on it.

STUDY QUESTIONS

WATER

1. All water on Earth is constantly in the process of being ________________ and ________________.

PHYSICAL STATES OF WATER

2. The three physical states of water are ________________, ________________, and ________________.

3. These three states are the result of different degrees of ________________ between water molecules.

4. What are the two forces that determine the different degrees of interaction between water molecules?

   a. ____________________ b. ____________________

5. Hydrogen bonding tends to (Attract, Separate) water molecules.

6. Kinetic energy tends to (Attract, Separate) water molecules.

7. Below freezing, (Hydrogen Bonding, Kinetic Energy) is the dominant force determining the association between water molecules together.

8. Above freezing, (Hydrogen Bonding, Kinetic Energy) is the dominant force determining the association between water molecules.

9. During **evaporation**, kinetic energy is (High, Low) when compared to hydrogen bonding.

10. During **condensation**, kinetic energy is (High, Low) when compared to hydrogen bonding.

11. Water going from a liquid to a vapor state is called ________________.

12. Water going from a vapor to a liquid state is called ________________.

EVAPORATION, CONDENSATION AND PURIFICATION

13. Evaporation and recondensation of water are an important purification process called ________________.

14. Is it (True, False) that all the natural freshwater on earth comes from this distillation process?

THE WATER CYCLE

15. The water cycle is also called the ________________ cycle.

WATER INTO THE ATMOSPHERE

16. Water vapor enters the atmosphere by way of:

    a. ______________________ from all water and moist surfaces.
    b. ______________________ which is really evaporation through the leaves of plants.

17. The amount of water vapor held in the air at various temperatures is called ______________ humidity.

18. Warm air holds (More, Less) water than cold air.

PRECIPITATION

19. Rising air results in (High, Low) precipitation whereas descending air results in (High, Low) precipitation.

20. Air rises over equatorial regions and descends over subequatorial regions, therefore

    a. rainfall is (High, Low) in equatorial regions.
    b. rainfall is (High, Low) in subequatorial regions.

21. The **rainshadow** refers to a region of (High, Low) precipitation.

WATER OVER AND THROUGH THE GROUND

22. When precipitation hits the ground, what two alternative pathways might it follow?

    a. ____________________ b. ____________________

23. All ponds, lakes, streams, and rivers are referred to as ______________ waters.

24. Water that infiltrates may follow either of two pathways including:

    a. ________________________ or b. ________________________ water.

25. Plants draw mainly from ________________________ water.

26. Water that percolates down through the soil eventually comes to an ______________________ layer and accumulates filling all the empty pores and spaces. This accumulated water is now called ____________________ water and its upper surface is called the ______________________ table.

27. The layers of porous material through which groundwater moves are called an ____________________.

28. Where water actually enters an aquifer is called the ______________ area.

29. Natural exits of groundwater to the surface are ______________ or ______________.

30. Springs and seeps feed streams and rivers becoming part of ______________ water.

SUMMARY OF THE WATER CYCLE

31. Summarize the below events in the water cycle by indicating whether they primarily occur as part of the [a] surface runoff loop, [b] evaporation-transpiration loop, or [c] groundwater loop.

    [ ] springs
    [ ] water table
    [ ] transpiration
    [ ] percolation
    [ ] surface runoff
    [ ] cloud formation
    [ ] water vapor
    [ ] capillary water
    [ ] gravitational water
    [ ] aquifers
    [ ] ground water
    [ ] infiltration

32. The process of material being dissolved and carried along by percolating water is called ______________.

HUMAN DEPENDENCE AND IMPACTS ON THE WATER CYCLE

33. Is it (True, False) that all water, in one way or another, comes through the water cycle?

34. Is it (True, False) that all wastes, in one way or another, are put into or at least exposed to the water cycle?

SOURCES AND USES OF FRESHWATER

35. Major trends in water use include

a. increasing demands for irrigation. (True, False)
b. increasing demands of municipal use (city water). (True, False)
c. conservation. (True, False)

36. Such use of water leads to problems such as

a. maintaining a constant supply of water. (True, False)
b. ecosystem damage. (True, False)
c. pollution. (True, False)

37. As water is used and reused it has the potential of becoming (More, Less) polluted.

38. When water is used, recycled, and reused, such use is called _________________ water.

39. When water is used and returned to the atmosphere, such use is called _________________ water.

40. Is it (True, False) that groundwater may be depleted by overuse?

41. Center pivot irrigation systems use huge amounts of (Surface, Ground) water and may use as much as (1, 10, 100) thousand gallons per minute.

CONSEQUENCES OF OVERDRAFT OF WATER RESOURCES

Overdraft of Surface Waters

42. No more than _________ percent of a river's average flow can be taken risking shortfalls.

43. However, the demand on some rivers exceeds ________ percent of the average flow.

44. A river's water being taken or diverted results in a decreased flow of fresh water along and to the mouth of that river. This may result in

a. wetlands along and at the mouth of the river drying up. (True, False)
b. estuaries at the mouth of the river becoming increasingly salty. (True, False)
c. dieoffs of large populations of wildlife. (True, False)

Overdraft of Groundwater

45. Depletion of ground water will very likely result in

a. cutbacks in agricultural production. (True, False)
b. diminishing surface water. (True, False)
c. adverse effects on stream and river ecosystems. (True, False)
d. certain land areas gradually sinking. (True, False)
e. formation of sinkholes. (True, False)
f. saltwater intrusion into wells in coastal regions. (True, False)
g. required cutbacks in municipal consumption. (True, False)

OBTAINING MORE VERSUS USING LESS

46. It may be possible to increase water supplies:

a. at modest cost. (True, False)
b. without severe ecological impacts. (True, False)

47. In the future it is very likely that water shortages will

a. become increasingly common. (True, False)
b. be of longer duration when they come. (True, False)
c. affect increasingly large areas and numbers of people. (True, False)
d. go away. (True, False)

POTENTIAL FOR CONSERVATION AND REUSE OF WATER

48. In low rainfall regions of the country, about _______ percent of the water is used in irrigation.

49. About _________ to ________ percent of irrigation water is wasted.

50. A water conserving alternative to surface and sprinkling irrigation is ______________ irrigation.

51. Water consumption in modern homes averages around ________ gallons per person per day.

52. The most effective conservation technique would be to use (More, The Same, Less) water.

53. An example of recycling water is ________________ water.

EFFECTS CAUSED BY CHANGING LAND USE

54. All the land draining into a particular stream or river is called its _________________.

55. As land is developed, natural soil surfaces are replaced by various hard surfaces. By this action

a. infiltration is (Increased, Decreased).
b. runoff is (Increased, Decreased).

56. As a result of the change in infiltration and runoff indicated above (#55) indicate whether each of the following will increase [+], decrease [-], or remain the same [0].

[ ] amount of groundwater recharge
[ ] level of water table
[ ] amount of flow from springs
[ ] stream flow between rains
[ ] stream flow during rains
[ ] stream bank erosion
[ ] undercutting and felling of trees
[ ] frequency of flooding
[ ] heights of floods

[ ] sediment deposition in the stream channel
[ ] breadth of stream channel
[ ] depth of stream channel
[ ] pollution of stream channel

57. List the categories of pollutants that enter streams and rivers from surface runoff.

a. ____________________, b. ____________________, c. ____________________,

d. ____________________, e. ____________________, f. ____________________,

g. ____________________, and h. ____________________.

58. ____________________ runoff is now recognized as the major source of pollution in many rivers.

59. Is it (True, False) that surface runoff recharges groundwater?

60. A stream that carries a large amount of water during rains, but which goes dry shortly after rains, is fed by

a. springs, b. runoff, c. there is no way of telling

61. As a region succumbs to development, the flow of streams in the area is likely to

a. remain unchanged, b. increase during rains, c. decrease between rains

STORMWATER MANAGEMENT

62. To alleviate the problem of flooding and stabilize the stream banks, streams may be **channelized.** In this process

a. a channel is dug out. (True, False)
b. curves are smoothed out. (True, False)
c. the channel is lined with concrete or rock. (True, False)
d. a natural water flow is restored. (True, False)
e. a natural stream ecology is restored. (True, False)

63. The modern ecological concept of stormwater management attempts to (Keep, Divert) the storm water and to (Preserve, Change) the natural infiltration/runoff ratio.

64. List five modern stormwater management techniques.

a. ____

b. ____

c. ____

d. ____

e. ____

IMPLEMENTING SOLUTIONS

CHEAP WATER AND WASTE

65. Current trends in water use (Are, Are Not) sustainable.

66. Is it (True, False) that people will not conserve that for which they do not pay a fair price?

KEY WORDS AND CONCEPTS

water quantity	water quality	evaporation	physical states of water
condensation	solution	suspension	fresh water
saltwater	polluted water	settling	purification of water
filtration	chemical absorption	water cycle	hydrological cycle
transpiration	precipitation	infiltration	surface runoff
groundwater	percolation	water table	groundwater recharge area
aquifer	leaching	spring	consumptive water use
water diversion	land subsidence	saltwater encroachment	nonconsumptive water use
drip irrigation	gray water	stream bank erosion	channelization
stormwater	watershed	watershed planning	stormwater management
stormwater reservoir			

SELF TEST

Circle the correct answer to each question.

1. Water will not exist as

   a. a solid, b. a liquid, c. separate oxygen and hydrogen atoms, d. a gas

2. The abiotic factor that influences the degree of interaction between water molecules is

   a. temperature, b. covalent bonding, c. hydrogen bonding, d. kinetic energy

3. Below freezing, what is the dominant force determining the association between water molecules?

   a. temperature, b. covalent bonding, c. hydrogen bonding, d. kinetic energy

4. The water cycle of the Earth is also called the

   a. Hydrogen Cycle.
   b. Hydrological Cycle.
   c. Hydronium Cycle.
   d. Hydrospheric Cycle.

5. Water going from a vapor to a liquid state is called

   a. evaporation, b. transpiration, c. condensation, d. percolation.

6. Water that accumulates underground above an impervious layer is called

   a. capillary water, b. gravitational water, c. ground water, d. an aquifer.

7. As air is warmed, its capacity to take up and hold more water is

   a. increased, b. decreased, c. remains constant, d. may change in unpredictable ways.

Match the below events of the water cycle by indicating whether they primarily occur as part of the [a. surface runoff loop, b. evaporation-transpiration loop, c. groundwater loop].

8. Water table recharge

9. Cloud formation

10. Capillary water

11. Surface runoff

12. Following the pathway of the water cycle, condensation leads most directly to

a. evaporation, b. precipitation, c. transpiration, d. infiltration

13. Following the pathway of the water cycle, capillary water leads most directly to

a. evaporation, b. precipitation, c. transpiration, d. infiltration

14. For the most part, plants draw on

a. ground water, b. capillary water, c. gravitational water, d. aquifers

15. Which of the following is not a major trend in water use?

a. Increasing demands for irrigation.
b. Increasing demands for municipal use.
c. Maintaining a constant water supply.
d. Damage of aquatic ecosystems.

16. Center pivot irrigation systems, for the most part, draw on

a. ground water, b. capillary water, c. gravitational water, d. surface water.

17. Depletion of ground water will very likely result in

a. cutbacks in agricultural production.
b. adverse effects on stream and river ecosystems.
c. formation of sinkholes.
d. all of the above

18. Which of the following predictions are not likely given future water shortages?

a. Water shortages will be averted through discoveries of new ground water resources.
b. Water shortages will become increasingly common.
c. Water shortages will be of a longer duration.
d. Water shortages will affect increasingly larger areas and more people.

19. As land is developed and natural soil surfaces are replaced by various hard surfaces

a. infiltration increases and run off decreases.
b. infiltration decreases and run off increases.
c. no changes in infiltration and run off rates occur.
d. there is no way of predicting the effect of development on either process.

20. Which of the following is not a result of decreasing infiltration and increasing surface run off?

a. The level of the water table will decrease.
b. Stream flow during rains will increase.
c. The depth of the stream channel will increase.
d. Water quality of the stream will decrease.

ANSWERS TO STUDY QUESTIONS

1. repurified, recycled; 2. solid, liquid, gas; 3. interaction; 4. hydrogen bonding, kinetic energy; 5. attract; 6. separate; 7. hydrogen bonding; 8. kinetic energy; 9. high; 10. low; 11. evaporation; 12. condensation; 13. distillation; 14. true; 15. hydrological; 16. evaporation, transpiration; 17. relative; 18. more; 19. high; 20. high, low; 21. low; 22. infiltration, runoff; 23. surface; 24. capillary, gravitational; 25. capillary; 26. impervious, ground, water; 27. aquifer; 28. recharge; 29. springs, seeps; 30. surface; 31. c, c, b, b, a, b, b, b, c, c, c, b; 32. leaching; 33. true; 34. true; 35. true, true, false; 36. all true; 37. more; 38. nonconsumptive; 39. consumptive; 40. true; 41. ground, 10; 42. 30; 43. 90; 44. all true; 45. all true; 46. all false; 47. true, true, true, false; 48. 85; 49. 50, 75; 50. drip; 51. 150; 52. less; 53. gray; 54. watershed; 55. decreased, increased; 56. -, -, -, -, +, +, +, +, +, +, +, -, -; 57. sediment, fertilizers, "cides", feces, salt, toxic chemicals, oil, litter; 59. true; 60. b; 61. b; 62. true, true, true, false, false; 63. keep, preserve; 64. dry wells and trenches, swales and barriers, parking lots with porous surfaces, retention ponds, capture of roof top and parking lot water; 65. are not; 66. false

---

ANSWERS TO SELF TEST

1. c; 2. a; 3. c; 4. b; 5. c; 6. a; 7. a; 8. c; 9. b; 10. b; 11. a; 12. b; 13. c; 14. b; 15. c; 16. a; 17. d; 18. a; 19. b; 20. c

CHAPTER 9

SEDIMENTS, NUTRIENTS, AND EUTROPHICATION

We all use the word **pollution** and we feel we know what it means. Yet, in trying to define it, we start to stumble, or our definition only covers one small aspect of the total. Indeed pollution constitutes so many different things that it defies simple definitions. For example, a simple definition such as "*the unfavorable alteration of our surroundings, wholly or largely as a by-product of man's action...*" is so broad and vague that it is virtually meaningless. The only way we can begin to understand the vastness and complexity of the pollution issue is to study the various kinds of things that constitute pollution. In this study, we should become increasingly aware that what we call pollution may be literally hundreds of different kinds of things from as many different sources.

You may want to begin thinking of pollutants in terms of *quantity* and *type*. Quantity is easier to understand than type. Often we expect natural ecosystems to "do something" with thousands of tons of ash, sediment, industrial wastes, and our garbage. Added to these demands, we also ask nature to decompose products that are unrecognizable (e.g., PVC pipe, toxic wastes and other nonbiodegradable materials) to decomposers. In effect, we are asking decomposers to "do their thing" without blueprints or road maps.

Likewise, stopping pollution is not simple. We must come to recognize that wastes are unavoidable by-products of civilization. There is no way that a civilization (or any other population of organisms) can exist without producing wastes. We cannot pretend that we can or will simply stop producing wastes. Instead, the solutions lie in learning to manage and recycle wastes so that they do not cause an "unfavorable alteration of our surroundings." In this chapter and the four that follow it, we will become familiar with major categories of wastes that arise from various sources and we will find that each kind of waste from each source requires its own control/management strategy.

STUDY QUESTIONS

EUTROPHICATION

TWO KINDS OF AQUATIC PLANTS

1. Understanding eutrophication requires recognizing two distinct kinds of aquatic plants: ________________________ plants which are rooted to the bottom and ________________________ aquatic vegetation which mostly consist of single cells or small groups of cells which grow at or near the water surface.

2. The depth to which light penetrates in sufficient intensity to support photosynthesis is known as the ________________ zone.

3. Depending on the turbidity cloudiness of the water, the euphotic zone may vary from nearly ______________ feet in clear water down to only a few inches in very turbid water.

4. Phytoplankton can maintain itself near the surface where there is plenty of light, but it depends on (Surface, Bottom) nutrients.

5. Benthic plants which are rooted in the bottom get their nutrients from the (Surface, Bottom).

6. Is it (True, False) that the balance between benthic plants and phytoplankton depends on the level of nutrients in the water?

7. Do (Benthic Plants, Phytoplankton) depend on light penetration to the bottom to enable photosynthesis?

UPSETTING THE BALANCE BY NUTRIENT ENRICHMENT

The Oligotrophic or Nutrient-poor Condition

8. The term **oligotrophic** means nutrient (Rich, Poor).

9. Nutrient-poor water (Limits, Promotes) the growth of phytoplankton but (Limits, Promotes) the growth of benthic plants.

Nutrient Enrichment, Eutrophication

10. The term **eutrophic** means nutrient (Rich, Poor).

11. The major source of oxygen in aquatic ecosystems comes from the

    a. atmosphere, b. benthic plants, c. phytoplankton

12. Phytoplankton do not contribute significantly to dissolved oxygen because

    a. it does not carry on photosynthesis. (True, False)
    b. it does not produce oxygen in photosynthesis. (True, False)
    c. the oxygen it produces escapes from the surface into the atmosphere. (True, False)

13. Does the amount of detritus (Increase, Decrease) in eutrophic water?

14. The increase in detritus comes from the dieback of (Benthic Plants, Phytoplankton).

15. Attack of detritus by ____________________ uses so much ______________________ that bottom dwelling fish and shellfish suffocate.

16. Decomposers (bacteria) can keep the dissolved oxygen at or near zero for as long as there is detritus to feed on because in the absence of oxygen they can carry on __________________ respiration.

17. Eutrophication includes

    a. nutrient enrichment. (True, False)
    b. depletion of benthic plants. (True, False)
    c. increase of phytoplankton. (True, False)
    d. increase in turbidity. (True, False)
    e. depletion of dissolved oxygen. (True, False)
    f. suffocation of fish and shellfish. (True, False)

18. A eutrophic body of water is not technically "dead" because it still supports the abundant growth of ________________ and ______________.

19. Indicate whether the following conditions would limit the growth of [a] benthic plants or [b] phytoplankton.

[ ] Nutrient-poor water.
[ ] Nutrient-rich water.
[ ] Clear water.
[ ] Turbid water.

20. A nutrient-poor body of water will most likely

a. have virtually no life. (True, False)
b. have clear water. (True, False)
c. have an ecosystem based on benthic plants. (True, False)
d. have a rich ecosystem with many species of fish and shellfish. (True, False)
e. be aesthetically pleasing for swimming, boating, and fishing. (True, False)

21. As a body of water becomes nutrient-rich, indicate whether the following changes would [+] or would not [-] occur.

[ ] Benthic plants would grow more abundantly.
[ ] Phytoplankton would grow more abundantly.
[ ] There would be increasing turbidity.
[ ] There would be decreasing turbidity.
[ ] There would be a greater diversity of aquatic plants and animals.
[ ] There would be less diversity of aquatic plants and animals.
[ ] There would be more oxygen.
[ ] There would be less oxygen.

NATURAL VERSUS CULTURAL EUTROPHICATION

22. Is it (True, False) that eutrophication of natural bodies of water is part of the natural aging process?

23. Cultural eutrophication is an (Accelerated, Abbreviated) form of natural eutrophication.

COMBATTING EUTROPHICATION

24. List the two general approaches to combatting eutrophication.

a. ____

b. ____

25. Indicate whether the following two approaches would be [a] attacking the symptoms or [b] getting at the root cause.

[ ] Reducing inputs of nutrients and sediments.
[ ] Getting rid of phytoplankton and increasing oxygen levels.

Attacking the Symptoms

26. List the three methods that have been used in attacking the symptoms of eutrophication and state the major reason that each is less that successful or practical. (see section subheadings)

	Method	Shortcoming
a.	______________	______________
b.	______________	______________
c.	______________	______________

Controlling Inputs

27. Is it (True, False) that the real control of eutrophication requires decreasing nutrient and sediment inputs?

SOURCES OF SEDIMENTS AND NUTRIENTS

SOURCES OF SEDIMENTS

28. The source of all sediments is ______________ erosion.

29. List seven major points of soil erosion and sediments.

a. ______________, b. ______________, c. ______________,

d. ______________, e. ______________, f. ______________,

g. ______________

IMPACTS OF SEDIMENTS ON STREAMS AND RIVERS

30. Indicate whether the following impacts of sediments on streams and rivers is the result of [a] sand, [b] silt, [c] clay and humus.

[ ] Deposits at the mouth of the river.
[ ] Settles out clogging the channel and slowing stream velocity.
[ ] Remains in suspension cutting off light and compounding eutrophication.

Damage to Aquatic Ecosystems of Streams and Rivers

31. Is it (True, False) that streams, rivers, lakes, and estuaries support complex ecosystems based on many kinds of plant and animal organisms living on or attached to the bottom?

32. Indicate whether the following types of damage to aquatic ecosystems is caused by [a] clay or [b] the bedload of sand and silt.

[ ] Reduction of light penetration and rate of photosynthesis.
[ ] Smothers fish by clogging their gills.
[ ] Scours bottom-dwelling organisms clinging to rocks.
[ ] Fills in hiding and resting places for fish and crayfish.
[ ] Causes the stream to become more shallow.

33. ____________________ is considered the foremost pollution problem of streams and rivers.

Filling of Channels and Reservoirs

34. Stream and river channels becoming clogged with sediments aggravate problems of ____________________ and ____________________.

35. Water supply reservoirs may be ____________________, shipping channels made ____________________, and irrigation canals are ____________________.

36. Specify the following problems associated with reliance on dredging.

a. economic ____________________

b. ecological ____________________

c. disposal ____________________

37. Is it (True, False) that dredging provides a permanent solution to sedimentation in streams and rivers?

38. The impacts of sediments cost the United States $_________ billion each year.

SOURCES OF NUTRIENTS

39. Nutrient ions such as ____________________, ____________________, and ____________________ cling to clay and humus.

40. List seven major sources of nutrients that end up in streams and rivers.

a. ____________________

b. ____________________

c. ____________________

d. ____________________

e. ____________________

f. ____________________

g. ____________________

LOSS OF WETLANDS AND BULKHEADING OF SHORELINES

41. Lands that are naturally covered by shallow water at certain times and more or less drained at others are called ____________________.

42. Marshes, swamps, and bogs are examples of ____________________.

43. Indicate whether the following are [+] or are not [-] functions of wetlands.

[ ] Water purification.
[ ] Groundwater recharge.
[ ] Filtering nutrients and sediments.
[ ] Provision of habitat and food for waterfowl and other wildlife.
[ ] Moderation of wave action.

44. Humans generally have a (Positive, Negative) attitude concerning wetlands.

45. Over _________ percent of the wetlands in the United States have been destroyed by draining or filling.

46. Indicate whether human development projects adjacent to or on wetlands increase [+] or decrease [-] the following wetland functions.

[ ] Water purification.
[ ] Groundwater recharge.
[ ] Filtering nutrients and sediments.
[ ] Provision of habitat and food for waterfowl and other wildlife.
[ ] Moderation of wave action.

CONTROLLING NUTRIENTS AND SEDIMENTS

47. Runoff and leaching from rural and urban areas is referred to as _______________ pollution.

48. Effluents from a specific source is called _________________ source pollution.

49. The cornerstone of federal water pollution control is the __________________ of 1972.

BEST MANAGEMENT PRACTICES ON FARMS, LAWNS, AND GARDENS

50. Provide an example of a best management practice under the following conditions.

a. Agricultural fields adjacent to streams, rivers, and lakes.

____

b. Feed lots, dairy barns, and horse stables near streams, rivers, and lakes.

____

c. Urban homes with expansive green lawns that are fertilized.

____

SEDIMENT CONTROL ON CONSTRUCTION AND MINING SITES

51. Is it (True, False) that construction and mining sites are major sources of sediments?

52. The sediment from a construction site may be controlled by constructing a _______________ trap.

53. Putting sediment traps on construction sites is an added expense. Therefore, erosion control on construction sites requires ________________ action and legal ________________.

PRESERVATION OF WETLANDS

54. Preservation of wetlands has (Short-term, Long-term, Both) benefits for humans and the environment.

BANNING THE USE OF PHOSPHATE DETERGENTS

55. Is it (True, False) that the detergent industry has developed substitutes for phosphate detergents?

56. Is it (True, False) that the detergent industry has stopped production of phosphate detergents?

57. Is it (True, False) that "phosphate bans" are now in effect in certain regions of the United States?

ADVANCED SEWAGE TREATMENT

58. Most sewage treatment plants (Do, Do Not) remove nutrients from waste water.

KEY WORDS AND CONCEPTS

pollution
biodegradable
nonbiodegradable
erosion
sediment
sediment deposit
sediment trap
photosynthesis
benthic plants
SAVs
turbidity
oligotrophic
dissolved nutrients
dissolved oxygen
eutrophication
herbicides
aeration
point source
nonpoint source
buffer strip
anaerobic respiration
phytoplankton
euphotic zone
nutrient enrichment

SELF TEST

Circle the correct answer to each question.

Indicate whether the following conditions would be indicative of [a] eutrophic or [b] oligotrophic aquatic ecosystems.

1. Nutrient rich

2. High light penetration

3. High species diversity

4. High decomposition rates

5. Which of the following approaches to combating eutrophication is getting at the root cause?

   a. chemical treatments, b. decreasing nutrient input, c. aeration, d. harvesting algae

6. Clay sediments in aquatic ecosystems

   a. reduce light penetration and the rate of photosynthesis.
   b. scour bottom-dwelling organisms clinging to rocks.
   c. fill in hiding places for fish and crayfish.
   d. cause the stream to become more shallow.

7. The foremost pollution problem of streams and rivers is

   a. toxic wastes, b. salinization, c. sedimentation, d. eutrophication

8. Reliance on dredging to clear streams of sediments results in

   a. high labor and machinery costs.
   b. destruction of the aquatic ecosystem.
   c. a problem of where to put the sediments.
   d. all of the above

9. The source of all sediments is

   a. erosion, b. leaching, c. weathering, d. feedlot dumps

10. Nutrient ions flow into aquatic ecosystems from

    a. fertilizers from croplands.
    b. animal wastes from feedlots.
    c. detergents containing phosphate.
    d. all of the above

11. Lands that are naturally covered by shallow water at certain times and more or less drained at others are called

    a. grasslands, b. wetlands, c. rain forests, d. estuaries

12. Functions of wetlands do not normally include

a. water purification.
b. provision of habitat and food for waterfowl and other wildlife.
c. provision of habitat for human development.
d. groundwater recharge.

13. Over _________ percent of the wetlands in the United States have been destroyed.

a. 20, b. 30, c. 40, d. 50

14. Which of the following wetland functions will increase after the establishment of human development projects adjacent to or on wetlands?

a. water purification, b. groundwater recharge, c. filtering, d. none of these will increase

15. Measures to ban the use of phosphate detergents have not included

a. development of substitutes for phosphate by the detergent industry.
b. enforcement of phosphate bans in certain regions of the United States.
c. the use of non-phosphate detergents by some of the U.S. public.
d. the total ban on phosphate detergent production.

ANSWERS TO STUDY QUESTIONS

1. benthic, submerged; 2. euphotic; 3. 100; 4. surface; 5. bottom; 6. true; 7. benthic plants; 8. poor; 9. limits, promotes; 10. rich; 11. a; 12. false, false, true; 13. increase; 14. benthic plants; 15. decomposers, oxygen; 16. anaerobic; 17. all true; 18. phytoplankton, fish; 19. a, b, a, b; 20. false, true, true, true, true; 21. -, +, +, -, -, +, -, +; 22. true; 23. accelerated; 24. attack the symptoms, get at the root cause; 25. b, a; 26. (chemical treatments, phytoplankton resistance), (aeration, expense), (harvesting algae, expense); 27. true; 28. soil; 29. croplands, overgrazed rangelands, deforested areas, construction sites, surface mining, gully erosion, miscellaneous; 30. b, a, c; 31. true; 32. a, a, b, b, b; 33. sediment; 34. erosion, flooding; 35. filled, impassable, clogged; 36. expense, destroys natural ecosystem, no where to put it; 37. false; 38. 6; 39. nitrate, phosphate, potassium; 40. fertilizer from cropland, fertilizer from lawns and gardens, animal wastes from feedlots, pet wastes, human excrement, phosphate detergents, acid precipitation; 41. wetlands; 42. wetlands; 43. all +; 44. negative; 45. 50; 46. all -; 47. nonpoint; 48. point; 49. Clean Water; 50. plant buffer strip of trees, construct collection ponds, compost organic wastes; 51. true; 52. sediment; 53. legislative, enforcement; 54. both; 55. true; 56. false; 57. true; 58. do not

---

ANSWERS TO SELF TEST

1. a; 2. b; 3. b; 4. a; 5. b; 6. a; 7. c; 8. d; 9. a; 10. d; 11. b; 12. c; 13. d; 14. d; 15. d

CHAPTER 10

WATER POLLUTION DUE TO SEWAGE

There is one pollutant that over the course of human history has caused hundreds of millions of deaths and innumerable cases of illness. This is hundreds of times more than the sum of all the other pollutants put together. This pollutant is now largely under control in developed countries, but it still causes millions of illnesses and many thousands of deaths each year in less developed countries. This most diabolical of all pollutants is untreated sewage.

In natural ecosystems, the breakdown of excrements and the reabsorption of the elements they contain is a crucial aspect of all the nutrient cycles. However, it is also possible to "recycle" various parasites and disease-causing organisms at the same time we are recycling the nutrients in excrements. This can lead to great epidemics of disease, suffering and death. Recall that in natural ecosystems such epidemics are a major factor in controlling populations and keeping them in balance with the rest of the ecosystem, but this kind of crude control we wish to avoid in human society.

The work of Louis Pasteur and others in the mid 1800s led us to recognize the disease hazards associated with excrements. Consequently, sewage collection and disposal systems were developed. These were the flush-away-with-water systems, the same concept that we still use today. These systems greatly reduce the disease hazard by removing the excrements from the immediate vicinity of human habitation. Thus, they reduce the potential of contaminating water and food supplies with the disease-causing organisms that may be present. Unfortunately, such systems also introduce large loads of nutrients into bodies of water. This transfer is a one-way street because only water evaporates and is recycled; nutrients remain behind as water evaporates. Even the most widely used sewage treatment processes do not remove the nutrients from the water. The consequence is that bodies of water are being increasingly overloaded with nutrient inputs, a consequence that is having growing ecological repercussions. This is to say nothing of the loss of nutrients, a valuable resource, from the land-based cycle.

It is now possible to have disease control, avoidance of pollution due to nutrient overloading, and recycling of nutrients. We shall proceed to learn about the process of sewage treatment and see how additions or alternatives can solve the problems.

STUDY QUESTIONS

HAZARDS OF UNTREATED SEWAGE

1. The three major problems resulting from untreated sewage wastes going into waterways are

   a. ______________________________ , b. ______________________________, and

   c. ______________________________ .

DISEASE HAZARD

2. Disease causing bacteria, viruses, and other parasitic organisms that infect humans and other animals are called ________________.

3. Transmission of pathogens in infected sewage waste may occur in what three ways?

a. ______________________, b. ______________________, c. ______________________

4. The level of infection of human populations (Is, Is Not) related to population density.

5. Public health measures which prevent a disease cycle involve

a. ____

b. ____

c. ____

DEPLETION OF DISSOLVED OXYGEN

6. Organic matter present in sewage that enter aquatic ecosystems

a. supports the growth of decomposers. (True, False)
b. leads to a depletion of dissolved oxygen. (True, False)

7. The concentration of sewage waste is expressed in terms of **BOD** or ______________________ ______________________ ______________________.

8. Indicate whether the following conditions would measure high [+] or low [-] in aquatic ecosystems showing high biological oxygen demand.

[ ] Oxygen concentration of the water.
[ ] The number of bacteria present.

9. Is it (True, False) that the depletion of dissolved oxygen will also suffocate the bacteria?

10. In order for bacteria to survive in an oxygen free environment, they must be able to carry on ______________________ metabolism.

EUTROPHICATION

11. Indicate whether the following conditions would measure high [+] or low [-] in aquatic ecosystems showing high biological oxygen demand.

[ ] Number of benthic plants.
[ ] Turbidity of the water.
[ ] Number of phytoplankton.
[ ] Level of phosphates, nitrates, and potassium.
[ ] Diversity of plants and animals.

SEWAGE HANDLING AND TREATMENT

12. Modern sewage treatment must address what three areas?

a. ______________________, b. ______________________, c. ______________________

BACKGROUND

13. Prior to the late-1800s, the general means of disposing of human excrements was the __________ privy which often was located near a __________.

14. What person showed the relation between sewage-borne bacteria and infectious diseases? __________

15. The earliest forms of sewage disposal transferred it directly into __________ drains.

16. __________ societies initiated the system of flushing sewage into natural waterways.

17. After the 1870s sewage treatment included __________ the wastewater and separating __________ and __________ water.

18. Is it (True, False) that there are no longer any cases of raw sewage overflowing with stormwater into waterways?

CONVENTIONAL SEWAGE TREATMENT

19. The total of all the drainage from sinks, bath tubs, laundry, and toilets constitutes a mixture called raw __________.

20. Raw sewage is actually __________ percent water and __________ percent pollutants.

21. List the three major categories of pollutants in raw sewage and give an example of each.

Category	Example
a. __________	__________
b. __________	__________
c. __________	__________

22. Match the below list of wastewater treatment technologies or products with [a] pretreatment, [b] primary treatment, [c] secondary treatment, or [d] advanced treatment.

[ ] primary clarifiers
[ ] bar screen
[ ] grit settling tank
[ ] raw sludge
[ ] biological treatment
[ ] trickling filter
[ ] activated sludge system
[ ] activated sludge
[ ] distillation or microfiltration
[ ] calcium phosphate
[ ] removal of dissolved nutrients
[ ] uses decomposers and detritus feeders
[ ] settling of colloidal material
[ ] settling of raw sludge
[ ] aeration

Pretreatment

23. First, large pieces of debris are removed by passing the water through a _______________ screen.

24. Coarse grit (sand and other dirt particles) are removed by passing the water through a ___________________ settling tank where the velocity of the water is (Increased, Decreased) and the particles are allowed to ___________________ out.

25. The large debris from the bar screen are taken out and _______________.

26. Grit from the grit settling tank is put in ___________________.

Primary Treatment

27. In primary treatment

    a. the water flows into very large tanks called _____________________ clarifiers.
    b. the water flows through (Quickly, Slowly).
    c. heavier particles of organic matter (Disperse, Settle Out).

28. What percentage of the colloidal material is removed by this settling process? _________

29. The material that settles out is called raw ______________________.

Secondary Treatment

30. Secondary treatment is also called ___________________ treatment because decomposers and detritus feeders are used.

31. The two types of secondary treatment systems used are:

    a. ____________________________ and b. _____________________________

32. In the trickling filter system, the water is allowed to trickle over ___________________ that are covered with a growth of ____________________ feeders attached to the rocks.

33. In the trickling filter system, it is the (Rocks, Organisms) that filter the wastewater.

34. After wastewater has passed through primary treatment and the trickling filter systems, __________ percent of the organic matter is removed.

35. In the activated sludge system, a mixture of organisms is added to the water to be treated and it is passed through a tank where it is ___________________.

36. In the aeration tank, the organisms feed on ________________ matter.

37. Following the aeration tank, the water is passed on to a secondary ___________________ tank where the organisms are returned to the _________________ tank.

38. Through the activated sludge systems, ___________ percent of the organic material is removed.

39. The most energy efficient system is the (Trickling Filter, Activated Sludge) system.

40. Is it (True, False) that the metabolism of organisms eventually releases nutrients into water?

41. After secondary treatment the water is generally chlorinated in order to kill ________________________.

Advanced Treatment

42. Nutrients may be removed by one of a number of methods referred to as ________________ treatment.

43. The main resistance to advanced treatment is ______________________ .

Disinfection

44. Indicate whether the following statements represent positive [+] or negative [-] attributes of using chlorine gas as the means of killing pathogenic bacteria in wastewater.

    [ ] Cost
    [ ] Toxicity during transportation and to fish.
    [ ] Chlorinated hydrocarbons.

45. Two alternative technologies for killing pathogenic bacteria in wastewater are

    a. ____________________ and b. ____________________

Sludge Treatment

46. Raw sludge is about ________ percent water and ________ percent organic matter.

47. Two methods of treating raw sludge to remove pathogenic bacteria and producing humus are

    a. ____________________ digesters and b. ____________________

48. Anaerobic digestion of raw sludge in ________________ produces the products of ________________ gas and ________________ sludge.

49. Composting basically produces ________________.

ALTERNATIVE SYSTEMS

50. List two alternative methods of using the nutrient-rich waste water.

    a. ______________________ and b. ____________________

TAKING STOCK OF WHERE WE ARE AND WHAT YOU CAN DO

PROGRESS AND LACK OF PROGRESS

51. Indicate whether the following statements indicate progress [+] or the lack of [-] progress in waste water treatment.

    [ ] Waste water treatment in less developed countries.
    [ ] Sewage being discharged with stormwater into natural waterways.
    [ ] Increasing population overburdening existing systems.
    [ ] Clean Water Act of 1972.
    [ ] Communities discharging raw sewage into bays, lakes, and coastal waters.
    [ ] Dumping sludge into landfills or the sea.

IMPEDIMENTS TO PROGRESS

52. List two major impediments to progress in waste water treatment.

    a. ______________________________

    b. ______________________________

53. Is it (True, False) that toxic chemical wastes can be removed through advanced treatment?

54. Is it (True, False) that people prefer contaminated drinking water than paying $2-3 more on their sewage bill?

WHAT YOU CAN DO

55. Is it (True, False) that you know how much treatment your sewage receives?

56. The fecal coliform test is an indirect method of monitoring sewage ______________.

57. A count of more than 200 *E. coli*

    a. is considered safe for swimming. (True, False)
    b. demonstrates a potential hazard. (True, False)
    c. demonstrates contamination due to human or other animal wastes. (True, False)

KEY WORDS AND CONCEPTS

disease hazards
parasites
dissolved oxygen
DO
biological oxygen demand
BOD
anaerobic respiration
eutrophication
suspended solids
organic colloidal material
detritus
dissolved material
biological treatment
trickling filters
activated sludge system
aeration tank
secondary clarifier
chlorination
sludge treatment
anaerobic digestion
biogas
methane
treated sludge
sludge cake
primary treatment
primary clarifier
raw sludge
secondary treatment
dry toilet
composting toilet
Clivus Multrum
E. coli
fecal coliform test

dissolved nutrients
pretreatment
bar screen
grit chamber, grit settling tank
humus
composting
toxic wastes
heavy metals

SELF TEST

Circle the correct answer to each question.

1. The major problems resulting from untreated sewage wastes going into waterways are

   a. disease, b. depletion of dissolved oxygen, c. eutrophication, d. all of these

2. The primary categories of pollutants in raw sewage are

   a. pathogens, detritus, and dissolved nutrients.
   b. phosphates, nitrates, and sulfates.
   c. heavy metals, synthetic organisms, and pesticides.
   d. industrial, household, and municipal wastes.

3. Public health measures which prevent the diseases associated with human wastes have not included

   a. disinfection of public water supplies.
   b. greater attention to personal hygiene and sanitation.
   c. public education on the personal dangers of sewage wastes.
   d. sanitary collection and treatment of sewage wastes.

4. Aquatic ecosystems that measure high biological oxygen demand would

   a. also register a high bacterial count.
   b. also register a high oxygen concentration.
   c. be crystal clear in appearance.
   d. contain a high diversity of benthic plants.

5. The role of modern sewage treatment plants is to control

   a. pathogens, b. B.O.D., c. nutrients, d. all of these

6. The first person to show the relation between sewage-borne bacteria and infectious diseases was

   a. Lewenhock, b. Redi, c. Pasteur, d. Aristotle

7. The earliest forms of sewage disposal

   a. separated out organic nutrients from waste water.
   b. did not separate waste water from storm water.
   c. were systems that used biological oxidation.
   d. used chlorine to disinfect waste water.

8. The waste water entering a sewage treatment plant is generally

a. 50% water and 50% polluting materials.
b. 70% water and 30% polluting materials.
c. 90% water and 10% polluting materials.
d. 99.9% water and 0.1% polluting materials.

Match each of the following facilities (a to d) with the function (questions 9 to 12) it performs.

a. activated sludge system, b. bar screen, c. primary clarifier, d. none of these

9. Screens out large pieces of debris.

10. Enables microorganisms to digest organic matter present.

11. Allows 50 to 70 percent of the organic matter to settle out.

12. Removes most of the dissolved nutrients.

13. In a sewage treatment plant, you see wastewater being sprayed on rocks with a giant sprinkler. This sprinkling system is part of

a. pretreatment, b. primary treatment, c. secondary treatment, d. advanced treatment

14. Most sewage treatment plants do not have

a. pretreatment, b. primary treatment, c. secondary treatment, d. advanced treatment

15. The treated water from most sewage treatment plants contains

a. nutrients, b. bacteria, c. sediments, d. colloidal materials

---

ANSWERS TO STUDY QUESTIONS

1. disease, depletion of dissolved oxygen, eutrophication; 2. pathogens; 3. drinking, eating food, body contact; 4. is; 5. disinfection of public water supplies, improving personal hygiene and sanitation, sanitary collection and treatment of sewage wastes; 6. all true; 7. biological oxygen demand; 8. -, +; 9. false; 10. anaerobic; 11. -, +, +, +, -; 12. pathogens, B.O.D., nutrients; 13. outdoor, stream; 14. Pasteur; 15. storm; 16. western; 17. treating, waste, storm; 18. false; 19. sewage; 20. 99.9, 0.1; 21. (debris and grit, rags and plastic bags), (colloidal or organic material, feces), (dissolved materials, nitrogen); 22. b, a, a, b, c, c, c, c, d, d, d, c, b, b, c; 23. bar; 24. grit, decreased, settles; 25. incinerated; 26. land fills; 27. primary, slowly, settles out; 28. 50 to 70; 29. sludge; 30. biological; 31. trickling filter, activated sludge system; 32. rocks, detritus; 33. organisms; 34. 85 to 90; 35. aerated; 36. organic; 37. clarifier, aeration; 38. 90 to 95; 39. trickling filter; 40. true; 41. pathogens; 42. advanced; 43. cost; 44. +, -, -; 45. ozone, radiation; 46. 98, 2; 47. anaerobic, composting; 48. digester, methane, treated; 49. humus; 50. irrigation, aquaculture; 51. -, -, -, +, -, -; 52. contamination with industrial wastes, public apathy; 53. false; 54. true; 55. false; 56. contamination; 57. false, false, true

---

ANSWERS TO SELF TEST

1. d; 2. a; 3. c; 4. a; 5. d; 6. c; 7. b; 8. d; 9. b; 10. a; 11. c; 12. d; 13. c; 14. d; 15. a

CHAPTER 11

TOXIC CHEMICALS AND GROUNDWATER POLLUTION

Before you begin studying Chapter 11, imagine someone approaching you and making the following proposal. This person says that he/she represents a major industry in the community that has provided hundreds of jobs to the residents. These jobs and the products produced by this industry have given economic security to the community. However, the industry is having some difficulty disposing of the toxic wastes resulting from product production. The industry is going to bury or discard the toxic wastes in an undetermined location with as little cost to the companies stockholders as possible. How would you react to this proposal? What questions do you need to have firm answers to before you would approve of it? Even though the above problem is occurring all over the United States, the way it has been presented to you is really a hypothetical situation. It is hypothetical because you would not be given the opportunity to react to the proposal but you would be given the opportunity to suffer the consequences.

Groundwater has traditionally been a major source of high quality water, suitable for drinking without further treatment or purification. Water supplies for about half the population of the United States are drawn directly from groundwater through personal or municipal wells. Assuming we keep withdrawal in balance with recharge groundwater could last forever, a renewable resource of immeasurable value. But, now groundwater supplies are presenting a new threat; with increasing frequency they are being found to be polluted with toxic chemicals which may have insidious health effects. These chemicals are not natural products. They are products and by-products of human activities. In short, we are poisoning our own well.

What are these chemicals? Where do they come from? What can we do to avoid contaminating ourselves with them? These are the questions which we will be addressing in this chapter.

STUDY QUESTIONS

1. Groundwater has been a traditional source of (High, Low) quality water which can be used for drinking with (Much, No) treatment or purification.

2. Is it (True, False) that there are increasing numbers of cases where groundwater is being found to be polluted?

SOURCES OF GROUNDWATER POLLUTION

3. The traditional high quality of groundwater is attributable to

   a. precipitation that is high quality fresh water. (True, False)
   b. filtration as water infiltrates and percolates through the soil. (True, False)
   c. soil bacteria degrading all natural organic waste products. (True, False)
   d. plants reabsorption of nutrients that are released in the breakdown to organic wastes. (True, False)

4. Is it (True, False) that groundwater is now being polluted by human-made chemical wastes which are toxic?

5. These materials are able to get into groundwater because they are (Biodegradable, Nonbiodegradable) and dissolve in water.

6. Significant sources of groundwater contamination (pollution) include:

   a. ______________________________

   b. ______________________________

   c. ______________________________

   d. ______________________________

   e. ______________________________

   f. ______________________________

   g. ______________________________

7. The most serious source of groundwater pollution is considered to be ______________.

TOXIC CHEMICALS: THEIR THREAT

WHAT ARE TOXIC CHEMICALS?

8. The two categories of toxic wastes that are particularly significant are ______________ and ______________.

Heavy Metals

9. Heavy metals include such metals as

   a. ____________ b. ____________ c. ____________

   d. ____________ e. ____________ f. ____________

   g. ____________ h. ____________

10. Heavy metals are extremely toxic because they combine with and inhibit the functioning of ______________ causing ______________ and ______________ effects.

Synthetic Organics

11. Synthetic organic chemicals are extremely toxic because they combine with and inhibit the functioning of ______________ causing ______________, ______________ and ______________ effects.

12. Synthetic organic chemicals are (Biodegradable, Nonbiodegradable).

13. Synthetic organic chemicals are the basis for the production of all ______________, synthetic ______________, synthetic ______________, ______________ coatings, ______________, ______________, and ______________ preservatives.

14. Indicate whether the following terms mean [a] cancer causing, [b] mutation causing, or [c] birth defect causing.

    [ ] mutagenic   [ ] carcinogenic   [ ] teratogenic

15. A particularly dangerous group of synthetic organic compounds are the ________________ hydrocarbons.

16. The main feature of halogenated hydrocarbons is that one or more hydrogen atoms have been substituted by an atom of ______________________, ______________________, ____________________ or ____________________.

17. By far the most common halogenated hydrocarbons are those containing chlorine. Such compounds are referred to as __________________ hydrocarbons.

18. Examples of chlorinated hydrocarbons include:

    a. __________________, b. __________________, c. __________________

    d. __________________, and e. ____________________.

THE PROBLEM OF BIOACCUMULATION

19. Both heavy metals and chlorinated hydrocarbons are particularly insidious because they tend to bio__________________________.

20. Bioaccumulation refers to the fact that small seemingly harmless doses of these compounds ingested with food or water over a long period of time (Increase, Decrease) in the body until harmful levels are reached.

21. Bioaccumulation of these compounds occurs because they are

    a. biodegradable. (True, False)
    b. nonbiodegradable. (True, False)
    c. readily absorbed by the body. (True, False)
    d. readily excreted from the body. (True, False)

22. Effectively, the body acts as a ______________________ for these compounds.

23. Bioaccumulation occurs

    a. just with humans.   b. with all organisms including humans.

24. When bioaccumulation is studied in the context of a food chain it is seen that each higher trophic (feeding) level receives and accumulates

    a. a higher,   b. a lower,   c. the same

    dose than the one before. (see figure 11-4)

25. Through a food chain the concentration of a toxic substance in organisms may be increased by as much as _______________ fold.

26. This concentrating effect that occurs through a food chain is called bio_______________.

27. Match [a] DDT, [b] PCBs, and [c] mercury with the bioaccumulation episode each caused.

[ ] Minamata Disease [ ] Diebacks of predatory birds [ ] Closing of fishing areas

SYNERGISTIC EFFECTS

28. When two or more substances interact together to cause effects that are far greater than the sum of the effect taken separately, it is known as a ____________________ effect.

ENVIRONMENTAL CONTAMINATION WITH TOXIC CHEMICALS

MAJOR SOURCES OF CHEMICAL WASTES

29. List four sources of chemical wastes.

a. ____________________

b. ____________________

c. ____________________

d. ____________________

BACKGROUND OF THE TOXIC WASTE PROBLEM

Indiscriminate Discharge into Air and Water

30. Prior to the passing of environmental laws in the 60s and 70s, it was common practice to dispose of chemical wastes into the ______________ or natural ______________.

31. Is it (True, False) that when a river catches fire society should become concerned about what is in the river?

32. The public outcry against pollution in the 60s led the U.S. Congress to pass two major pieces of legislation, namely the

a. ______________________________ Date ____________

b. ______________________________ Date ____________

33. In the years since these laws were passed direct discharges of wastes into natural waterways and the air have (Decreased, Increased).

Shift to Land Disposal

34. In order to diminish direct discharges of wastes into air and water, industries generally turned to

a. not producing them b. land disposal

35. This shift to land disposal of chemical wastes (Increased, Decreased) the potential for groundwater pollution.

METHODS OF LAND DISPOSAL

36. Indicate whether the following statements describe the disposal technique of [a] deep well injection, [b] surface impoundments, or [c] landfills.

    [ ] Concentrated wastes put in drums and buried.
    [ ] A reverse well.
    [ ] 57% of hazardous wastes disposed of in this manner.
    [ ] An open pond without an outlet.
    [ ] Volatile wastes may enter the atmosphere.
    [ ] 38% of hazardous wastes disposed of in this manner.
    [ ] Waste put into deep rock strata.
    [ ] Involves the removal of leachate.

37. Is it (True, False) that when these facilities are properly sited, constructed, and maintained, they guarantee that wastes will never seep or leach into groundwater?

PROBLEMS IN MANAGING LAND DISPOSAL

38. Two problems inherent in land disposal is that wastes ________________ at disposal facilities and that they ____________________ there.

39. The mysterious appearance of drums of toxic wastes in warehouses, vacant lots, and municipal landfills is probably the result of ____________________ dumping.

40. Another problem is the use of designated disposal sites that have no precautions against ground pollution. These sites are called __________________ facilities.

41. Love Canal is a classic example of

    a. midnight dumping. b. use of nonsecure facilities.

SCOPE OF THE PROBLEM

42. There are an estimated ________________ million tons of hazardous wastes generated each year in the United States.

43. Indicate whether the following problems of improper disposal pertain to [a] deep well injection, [b] surface impoundments, [c] landfills, or [d] all three.

    [ ] Absence of liners or leachate collection systems.
    [ ] Deposition of wastes above or into aquifers.
    [ ] Immersion of wastes into ground water.
    [ ] Lack of reliable monitoring systems to detect leakage.

44. Is it (True, False) that once wastes have seeped into groundwater, there is no practical way to remove them?

45. Small amounts of toxic chemicals in groundwater are extremely hazardous because of their known ability to

    a. bioaccumulate. (True, False)
    b. cause serious health problems at low concentrations. (True, False)

CLEANUP AND MANAGEMENT OF TOXIC WASTES

46. List the four major aspects to the toxic waste problem.

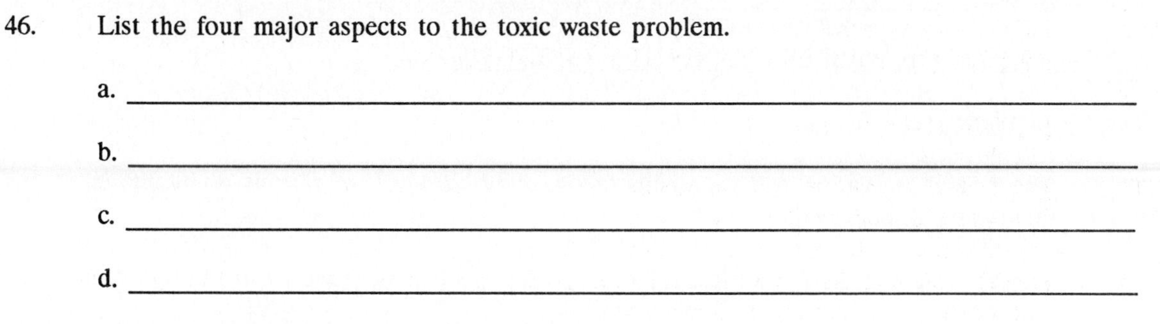

a. ____________________

b. ____________________

c. ____________________

d. ____________________

ASSURING SAFE DRINKING WATER SUPPLIES

47. A federal law that attempts to assure safe drinking water supplies is the ____________________ Date __________

48. Indicate whether the following are [+] or are not [-] provisions of the Safe Drinking Water Act of 1974.

[ ] Setting standards regarding allowable levels of pollution.
[ ] Monitoring municipal water supplies.
[ ] Proper location and construction of injection wells.
[ ] Systematic monitoring of groundwater and private wells.
[ ] Determination of "safe" and "unsafe" levels for all chemicals.
[ ] Provision of funds to compensate for illness resulting from contaminated wells.

CLEANING UP EXISTING TOXIC WASTE SITES

49. A federal program aimed at identifying and cleaning up existing waste sites is the

____________________ of 1980.

50. The Comprehensive Environmental Response, Compensation, and Liability Act of 1980 is commonly known as the ____________________.

51. The Superfund for 1980 to 1985 was $__________ billion.

52. The Superfund 1986 to 1991 is $__________ billion.

53. The administrative unit for the Superfund is the ____________________.

54. Indicate whether the following have been positive [+] or negative [-] attributes of the EPA's administration of the Superfund over the past five years.

[ ] Placement of sites on the "superfund list".
[ ] Scope of action across the United States.
[ ] Transport of toxic wastes from one site to another site.

GROUNDWATER REMEDIATION

55. Is it (True, False) that once groundwater is contaminated, it is effectively lost forever?

56. The process of pumping out contaminated water, passing it through filters, reinjecting it is called ____________________ remediation.

MANAGEMENT OF WASTES CURRENTLY PRODUCED

The Clean Water Act

57. The federal legislation that set standards on allowable levels of various pollutants in wastewater discharges is ________________ ____________________ Act.

58. Indicate whether the following have been positive [+] or negative [-] attributes of the administration of the Clean Water Act.

[ ] Interim permits
[ ] Monitoring system
[ ] Paying fines
[ ] Exemptions

59. Is it (True, False) that the Clean Water Act has stopped nearly all dumping of toxic wastes into natural waterways?

The Resources Conservation and Recovery Act

60. The law that addresses the management of hazardous wastes currently being produced is the RCRA or ____ of 1976.

61. The RCRA legislation required

a. records on amounts and kinds of all hazardous wastes. (True, False)
b. location of point of origin and disposal. (True, False)
c. standards of production and operation of waste disposal facilities. (True, False)

62. The RCRA legislation resulted in

a. the closure of many waste disposal facilities. (True, False)
b. disposal of hazardous wastes in inadequate facilities. (True, False)
c. exemptions from the law. (True, False)
d. interim permits. (True, False)
e. disposal of toxic wastes down sewers. (True, False)
f. tougher laws in some states. (True, False)
g. shipment of hazardous wastes to other countries. (True, False)
h. wholesale movement of some industries to other countries. (True, False)

FUTURE MANAGEMENT OF HAZARDOUS WASTES

63. List three ways in which the future management of hazardous wastes may be improved.

a. ____

b. ____

c. ____

64. Indicate whether the following treatments of hazardous wastes pertain to

[a] waste reduction, reclamation, and recycling
[b] incineration
[c] biodegradation

[ ] Maintaining waste inventories.
[ ] Biological breakdown of hazardous wastes to harmless by-products.
[ ] Abiotic breakdown of hazardous wastes to harmless by-products.

REDUCING OCCUPATIONAL AND ACCIDENTAL EXPOSURES

Right-to Know Legislation

65. Working with and around toxic materials may result in ________________ or ________________ exposures.

66. Title III of the Superfund Amendments and Reauthorization Act of 1986 is commonly known as "____________________" legislation.

67. Is it (True, False) that "right-to-know" legislation made it mandatory for employers to provide written information to their employees on the nature and hazards of materials in the work environment?

68. Is it (True, False) that "right-to-know" legislation made it mandatory for employers to make their employees read printed information on the nature and hazards of materials in the work environment?

Accidents and Emergency Response

69. Give examples of the toxic waste disasters that occurred in the following locations.

a. Bhopal, India 1984

____

b. Michigan 1973

____

c. Times Beach, Missouri

____

70. Over the course of a year, about ________ billion tons of hazardous materials are transported on the roads, railroad tracks, and waterways of the United States.

71. List one response of the following groups for preventing or reacting to toxic waste accidents.

a. Department of Transportation

____

b. Emergency response teams

c. Society in general

d. You personally

KEY WORDS AND CONCEPTS

leaching
groundwater
toxic
heavy metals
synthetic organic chemicals
carcinogenic
halogen
halogenated hydrocarbon
chlorinated hydrocarbon
bioaccumulation
biomagnification
synergistic effect
deep well injection
surface impoundment
landfill
nonsecure disposal
RCRA
teratogenic
mutagenic
Minamata disease
Clean Water Act 1972
Clean Air Act 1972
leachate
secure landfill
EPA
groundwater remediation
interim permits
waste inventories
biodegradation
SARA, Title III
emergency response teams
midnight dumping
Superfund

SELF TEST

Circle the correct answer to each question.

1. The traditional high quality of groundwater is the result of

   a. precipitation that is high quality freshwater.
   b. filtration of water as it infiltrates through the soil.
   c. decomposition of organic wastes by soil bacteria.
   d. all of the above

2. The most serious source of groundwater pollution is

   a. inadequate landfills.
   b. pesticides.
   c. sewage.
   d. transportation spill.

3. The two categories of toxic wastes that are particularly significant are

a. midnight dumping and inadequate landfills.
b. heavy metals and synthetic organics.
c. deicing salt and waste road oil.
d. sewage sludge and transportation spills.

4. Which of the following is not a heavy metal?

a. tin, b. lead, c. mercury, d. aluminum

5. Heavy metals and synthetic organics both cause

a. physiological and neurological effects.
b. mutagenic and carcinogenic effects.
c. teratogenic and mutagenic effects.
d. all of the above

6. Synthetic organics are known to cause

a. mutagenic effects, b. carcinogenic effects, c. teratogenic effects, d. all of these effects

7. A birth defect causing effect is termed

a. mutagenic, b. neurological, c. carcinogenic, d. teratogenic

8. Which of the following atoms is not a halogen?

a. chlorine, b. bromine, c. aspergine, d. iodine

9. Which of the following is not an example of a chlorinated hydrocarbon product?

a. pesticides, b. fertilizers, c. electrical insulation, d. plastics

10. A compound that has the quality of bioaccumulation

a. is biodegradable.
b. is readily excreted from the body.
c. will increase in concentration at each trophic level.
d. all of the above are qualities

11. DDT is added to a lake ecosystem. In which of the following organisms would one find the highest concentrations of DDT?

a. water, b. phytoplankton, c. zooplankton, d. fish

12. Minamata disease is the result of bioaccumulating

a. chlorinated hydrocarbons, b. heavy metals, c. pesticides, d. DDT

13. When two or more substances interact together to cause effects that are far greater than the sum of the effect taken separately, this is known as

a. bioaccumulation, b. biomagnification, c. a synergistic effect, d. a carcinogenic effect

14. Fifty-seven percent of hazardous wastes are removed through

    a. deep well injection, b. surface impoundments, c. landfills, d. all of these

15. Volatile wastes may enter the atmosphere from

    a. deep well injection, b. surface impoundments, c. landfills, d. all of these

16. Lack of monitoring systems to detect toxic waste leakage is a problem of

    a. deep well injection, b. surface impoundments, c. landfills, d. all of these

17. Midnight dumping relates to the specific toxic waste problem of

    a. cleaning up existing toxic waste sites.
    b. cleaning up contaminated ground water.
    c. proper disposal of existing toxic wastes.
    d. assuring safe drinking water.

18. The Safe Drinking Water Act of 1974 does not provide for

    a. standards regarding allowable levels of pollution.
    b. the monitoring of municipal water supplies.
    c. the proper location and construction of injection wells.
    d. systematic monitoring of groundwater and private wells.

19. The federal program aimed at identifying and cleaning up existing waste sites is the

    a. Resources Conservation and Recovery Act of 1976.
    b. Clean Water Act of 1974.
    c. SARA, Title III of 1986.
    d. Comprehensive Environmental Response, Compensation, and Liability Act of 1980.

20. The federal law that addresses the management of hazardous wastes currently being produced is

    a. Resources Conservation and Recovery Act of 1976.
    b. Clean Water Act of 1974.
    c. SARA, Title III of 1986.
    d. Comprehensive Environmental Response, Compensation, and Liability Act of 1980.

ANSWERS TO STUDY QUESTIONS

1. high, no; 2. true; 3. all true; 4. true; 5. biodegradable; 6. inadequate landfills, leaking underground storage tanks, pesticides and fertilizers, deicing salt, waste oil, sewage sludge, transportation spills; 7. pesticides; 8. heavy metals, synthetic organics; 9. lead, mercury, arsenic, cadmium, tin, chromium, zinc, copper; 10. enzymes, physiological, neurological; 11. enzymes, mutagenic, carcinogenic, teratogenic; 12. nonbiodegradable; 13. plastics, fibers, rubber, paint, solvents, pesticides, wood; 14. b, a, c; 15. halogenated; 16. chlorine, bromine, fluorine, iodine; 17. chlorinated; 18. plastics, pesticides, solvents, electric insulation, flame retardants; 19. accumulate; 20. increase; 21. false, true, true, false; 22. filter; 23. b; 24. a; 25. 100,000; 26. magnification; 27. c, a, b; 28. synergistic; 29. leftovers from chemical processes, used processing and cleaning materials, wash wastes from cleaning finished products, residues remaining in empty containers; 30. sewers, waterways; 31. true; 32. (Clean Air Act, 1970), (Clean Water Act, 1972); 33. decreased; 34. b; 35. increased; 36. c, a, a, b, b, b, a, c; 37. false; 38. arrive, stay; 39. midnight; 40. nonsecure; 41. b; 42. 150; 43. c, a, d, d; 44. true; 45. true, true; 46. assure safe water, clean-up existing sites, clean-up contaminated ground water, proper disposal of existing wastes; 47. Safe Drinking Water Act, 1974; 48. +, +, +, -, -, -; 49. Comprehensive Environmental Response, Compensation, and Liability Act; 50. superfund; 51. 1.6; 52. 8.6; 53. EPA; 54. all -; 55. false; 56. groundwater; 57. Clean Water; 58. all -; 59. false; 60. Resources Conservation and Recovery Act; 61. all true; 62. all true; 63. waste reduction, incineration, biodegradation; 64. a, c, b; 65. occupational, accidental; 66. right-to-know; 67. true; 68. false; 69. leak of methyl isocyanate, PCB contaminated livestock feed, dioxin contamination of road bed; 70. 4; 71. regulations of inter-state transport of toxic wastes, specialized training to respond to toxic waste accidents, demand responsible management, become informed

ANSWERS TO SELF TEST

1. d; 2. b; 3. b; 4. d; 5. a; 6. d; 7. d; 8. c; 9. b; 10. c; 11. d; 12. b; 13. c; 14. a; 15. b; 16. d; 17. c; 18. d; 19. d; 20. a

CHAPTER 12

AIR POLLUTION AND ITS CONTROL

Natural forest fires! Volcanic eruptions! Dust storms! Such events have always thrown foreign particles and gases (pollutants) into the atmosphere. But, there are also natural purification processes at work. Larger particles settle of their own accord and smaller particles and water soluble gases are continually cleansed from the atmosphere by the water cycle. Water vapor condenses on particles and gases may dissolve in the forming droplet and, thus, both are brought down with precipitation. Some foreign gases may be absorbed and assimilated by vegetation and soil microorganisms and if the amounts are not too great, the organisms do not suffer. Since natural events causing air pollution are generally few and infrequent, the natural biosphere maintained a balance between pollution input and removal in which the balance generally rested on clean air.

But now each year in the United States alone, we humans burn some 5 billion barrels of oil, 1 billion tons of coal, 22 trillion cubic feet of natural gas, mountains of refuse, and then add to this the materials from metal refining, all the fumes from organic chemicals that are produced in their manufacture and use. What happens to the balance? Indeed, that the air is not much worse than it is speaks to the remarkable capacity of natural processes and some to our own efforts in pollution control. However, should we be satisfied and complacent? Forests in many areas are dying indicating overburdening, too much stress from pollution, and at many times and places, pollution indexes reach unhealthful if not hazardous levels. Clearly, vigilance is in order and we need to do more to curtail our emissions of pollutants into the atmosphere. This chapter will describe what the major air pollutants are, where they come from, and what is being done and remains to be done to control them.

STUDY QUESTIONS

BACKGROUND OF THE AIR POLLUTION PROBLEM

1. Is it (True, False) that organisms do have the capacity to deal with certain amounts or levels of pollutants without suffering ill effects?

2. The level of pollutant below which no ill effects are observed is called the ________________ level.

3. Is it (True, False) that there are some compounds that have no threshold level?

4. Dose is defined as ________________ multiplied by time of ________________.

5. Which of the below factors determine the level of pollution?

   a. inputs, b. dispersal, c. out-takes, d. all of these

6. Is it (True, False) that air pollution began to occur only in the 20th century?

7. The discovery of ____________ began human inputs of air pollutants.

8. A common misconception among humans was that "the solution to pollution is ________________".

9. Thousands of cars venting unprotected exhaust + sunlight + a mountain topography = __________________ smog.

10. Normally temperature (Increases, Decreases) as elevation increases.

11. The warmer surface air ________________ which tends to carry pollutants away.

12. Warmer surface air cannot rise when a layer of (Warmer, Cooler) air overlies cooler air, a situation that is referred to as a ______________________ inversion.

13. Air pollution may

    a. adversely affect human health. (True, False)
    b. cause damage to crops and forests. (True, False)
    c. increase the rate of corrosion and deterioration of materials. (True, False)

14. Is it (True, False) that human illnesses and damage to crops and forests from air pollution has become increasingly commonplace?

15. A major law passed by the U.S. Congress which addresses the problem of air pollution is the ____________________ of __________ and its amendments of __________.

16. The standards of the Clean Air Act of 1970

    a. set the levels of pollution control needed to protect the environment and human health. (True, False)
    b. set the time tables for meeting pollution control levels. (True, False)

17. Identify the four steps involved in meeting the mandates of the Clean Air Act of 1970.

    a. ____

    b. ____

    c. ____

    d. ____

18. Is it (True, False) that air quality in most cities is better now than it was in the early 1970's?

MAJOR POLLUTANTS AND THEIR EFFECTS

MAJOR POLLUTANTS

19. List the eight pollutants or pollutant categories that are the most widespread and serious.

    a. ____________________ b. ____________________ c. ____________________

    d. ____________________ e. ____________________ f. ____________________

    g. ____________________ h. ____________________

20. In large measure, all of the above pollutants are direct or indirect products of _______________ (See Figure 12-7).

ADVERSE EFFECTS OF AIR POLLUTION ON HUMANS, PLANTS, AND MATERIALS

21. Air pollution

a. is the result of one chemical mixed with the normal constituents of air. (True, False)
b. varies in time and place. (True, False)
c. varies with environmental conditions. (True, False)

22. Is it (Easy, Difficult) to determine the role of particular pollutants in causing an observed result?

Effects on Human Health

23. Is it (True, False) that there is a clear and proven relationship between air pollution and human sickness or death?

24. Serious adverse effects of air pollution on human health are seen

a. mainly among smokers, b. mainly among nonsmokers, c. equally among both groups

25. Is it (True, False) that smoking by itself greatly increases the risk of serious disease?

26. Learning disabilities in children and high blood pressure in adults are correlated with high levels of _______________ in the blood.

27. The source of this widespread contamination is

a. people who smoke, b. use of leaded gasoline, c. high ozone levels, d. pesticide use

28. The EPA mandated the phase-out of leaded gasoline by the end of 1988. Is it (True, False) that you can still buy leaded gasoline in 1989?

Effects on Agriculture and Forests

29. Air pollution

a. may destroy vegetation. (True, False)
b. may seriously retard the growth of crops and forests. (True, False)
c. may cause forests to become more vulnerable to insect pests and disease. (True, False)

30. Indicate whether the following effects of air pollutants were [+] or were not [-] demonstrated in open-top chamber experiments.

[ ] Which pollutants cause the damage.
[ ] The sensitivity level of plants to gaseous air pollutants.
[ ] The degree to which air pollutants retard plant growth and development.
[ ] The degree to which air pollutants reduce crop yields and profits.

31. The most serious pollutants on agriculture and forests are _______________ and _______________ rain.

32. Is it (True, False) that all plants react in the same manner to air pollutants?

33. Is it (True, False) that just a small increase in concentration or duration of exposure may push some plants beyond their ability to cope with an air pollutant?

34. The point of intolerance to an air pollutant is called the ________________ level.

Effects on Materials and Aesthetics

35. Indicate whether the following conditions could [+] or could not [-] be an effect of air pollution.

[ ] Grey and dingy walls and windows.
[ ] Deteriorated paint and fabrics.
[ ] Corrosion of metals.
[ ] A bright, clear, blue sky.
[ ] Increased real estate values.
[ ] Decreased real estate values.

SOURCES OF POLLUTANTS AND CONTROL STRATEGIES

POLLUTANTS: PRODUCTS OF COMBUSTION

36. Air pollutants are direct or indirect by-products from burning ________________, ________________, and ________________.

37. Oxidation of coal, gasoline, and refuse by burning (Is, Is Not) usually complete.

38. Indicate whether the following are direct [D] or indirect [I] products (pollutants) from combustion.

[ ] particulates
[ ] hydrocarbon emissions
[ ] carbon monoxide
[ ] nitric oxide
[ ] sulfur dioxide
[ ] ozone
[ ] photochemical oxidants
[ ] sulfuric acid
[ ] nitric acid

SETTING STANDARDS

39. The law that mandated setting standards for air pollutants is the ________________.

40. The Clean Air Act mandated setting standards for

a. all air pollutants. (True, False)
b. five pollutants recognized as most widespread and objectionable. (True, False)

41. These five pollutants are known as ________________ pollutants.

42. List the five criteria pollutants.

a. ________________, b. ________________, c. ________________

d. ________________, e. ________________.

43. The highest level of a pollutant that can be tolerated by humans without noticing any ill effects plus a 10 to 50 percent margin of safety is called the __________________ standard.

44. Good air quality implies that the Pollution Standard Index (PSI) is within _________ percent of its Standard.

45. A PSI of _________ percent over the standard indicates air quality is in the "unhealthful" range.

MAJOR SOURCES AND CONTROL STRATEGIES

Indicate whether the strategies described below were designed to control emissions of

a. particulates
b. nitrogen oxides, hydrocarbons, ozone, and carbon monoxide
c. sulfur dioxide and acids
d. lead and other heavy metals

46. [ ] Phasing out the use of leaded gasoline.

47. [ ] Building higher smoke stacks.

48. [ ] Putting catalytic converters on car exhaust systems.

49. [ ] Engine adjustments that reduced combustion temperatures and pressures.

50. [ ] Installation of filters and electrostatic precipitators at industrial sites.

51. [ ] Closing of open burning dumps.

52. [ ] Banning of open burning of trash and refuse in cities.

53. For which of the above pollutants (a, b, c, d) has industry been made the most **un**accountable for emission control?

54. Which of the above pollutants (a, b, c, d) is a major source of acid precipitation?

TAKING STOCK - WHAT YOU CAN DO

55. List five improvements that can be added to the next reauthorization of the Clean Air Act.

a. ____

b. ____

c. ____

d. ____

e. ____

INDOOR AIR POLLUTION

56. Is it (True, False) that air inside the home and workplace often contains much higher levels of pollutants than outdoor air?

57. Factors contributing to indoor air pollution include

    a. products used indoors. (True, False)
    b. well insulated and sealed buildings. (True, False)
    c. duration of exposure to indoor pollutants. (True, False)

58. List five types or sources of indoor air pollutants.

    a. ____________________, b. ____________________, c. ____________________

    d. ____________________, e. ____________________

KEY WORDS AND CONCEPTS

threshold level	dose	photochemical smog
temperature inversion	air pollution disasters	indoor pollutants
Clean Air Act 1970, 1977	particulates	hydrocarbons
carbon monoxide	nitrogen oxides	sulfur oxides
heavy metals	ozone	photochemical oxidants
acids	synergistic effect	lung disease
lead poisoning	open-chamber experiments	direct products
indirect products	setting standards	primary standard
criteria pollutants	precipitators	catalytic converters
acid precipitation		

SELF TEST

Circle the correct answer to each question.

1. Which of the following statements is not accurate concerning the air pollution problem?

    a. Organisms have the capacity to deal with certain amounts and levels of pollution.
    b. There is a threshold pollution level that is tolerable.
    c. Air pollution has been with human society since the discovery of fire.
    d. The solution to pollution is dilution.

2. Unprotected car exhaust + sunlight + a mountain topography =

    a. a thermal inversion, b. photochemical smog, c. an air pollution episode, d. acid rain

3. The condition of a warm air mass overlying a cool air mass is referred to as

    a. a thermal inversion, b. photochemical smog, c. an air pollution episode, d. acid rain

4. Air pollution may

   a. adversely affect human health.
   b. cause damage to crops and forests.
   c. increase the rate of corrosion and deterioration of materials.
   d. cause all of the above effects.

5. A major problem with the Clean Air Act of 1970 and its 1977 amendment is

   a. identifying pollutants.
   b. demonstrating cause and effect relationships.
   c. determining the source of pollutants.
   d. developing and implementing controls.

6. Which of the following is not a major pollutant?

   a. particulates, b. hydrocarbons, c. carbon dioxide, d. nitrogen oxide

7. A major problem with air pollution control is

   a. identifying pollutants.
   b. demonstrating cause and effect relationships.
   c. determining the source of pollutants.
   d. developing and implementing controls.

8. Air pollution is known to have adverse effects on

   a. human health, b. agricultural crops, c. forests, d. all of these

9. Which of the following is considered the most serious pollutant on agriculture crops and forests?

   a. hydrocarbons, b. ozone, c. heavy metals, d. carbon monoxide

10. Learning disabilities in children and high blood pressure in adults are correlated with high blood levels of which of the following pollutants?

    a. hydrocarbons, b. ozone, c. heavy metals, d. carbon monoxide

    Indicate whether the following pollutants are [a] direct or [b] indirect products of combustion.

11. Particulates

12. Sulfuric acid

13. Ozone

    Indicate whether strategies 14 and 15 were designed to control emissions of

    a. particulates
    b. nitrogen oxides, hydrocarbons, and ozone
    c. sulfur dioxide and acids
    d. lead and other heavy metals

14. Building higher smoke stacks.

15. Putting catalytic converters on car exhaust systems.

ANSWERS TO STUDY QUESTIONS

1. true; 2. threshold; 3. true; 4. concentration, exposure; 5. d; 6. false; 7. fire; 8. dilution; 9. photochemical; 10. decreases; 11. rises; 12. warmer, thermal; 13. all true; 14. true; 15. Clean Air Act, 1970, 1977; 16. all true; 17. identifying pollutants, demonstrating cause and effect relationships, determining source, developing and implementing controls; 18. true; 19. particulates, hydrocarbons, carbon monoxide, nitrogen oxides, sulfur oxides, heavy metals, ozone, acids; 20. combustion; 21. all true; 22. difficult; 23. true; 24. a; 25. true; 26. lead; 27. b; 28. true; 29. all true; 30. all +; 31. ozone, acid; 32. false; 33. true; 34. critical; 35. +, +, +, -, -, +; 36. coal, gasoline, refuse; 37. is not; 38. d, d, d, d, d, i, i, i, i; 39. Clean Air Act of 1970; 40. false, true; 41. criteria; 42. particulates, sulfur dioxide, ozone, carbon monoxide, nitrogen oxides; 43. primary; 44. 25; 45. 150; 46. d; 47. c; 48. b; 49. b; 50. a; 51. a; 52. a; 53. c; 54. c; 55. set and enforce standards, control nitrogen oxides, apply standards to trucks and buses, address nonvehicle sources of hydrocarbons, reduce sulfur dioxide emissions; 56. true; 57. all true; 58. formaldehyde, cleaners, pesticides, aerosol sprays, radon

ANSWERS TO SELF TEST

1. d; 2. b; 3. a; 4. d; 5. b; 6. a; 7. b; 8. d; 9. b; 10. c; 11. a; 12. b; 13. b; 14. c; 15. d

CHAPTER 13

ACID PRECIPITATION, THE CO_2 GREENHOUSE EFFECT, AND DEPLETION OF THE OZONE SHIELD

Chapter 13 focuses on the three pollution problems that threaten all life on Earth. The last chapter provided an overview of the variety of pollutants the human ecosystem expects the natural ecosystem to dilute, assimilate or to just make it go away! This chapter describes the peril and promise as the human ecosystem attempts to control acid precipitation, carbon dioxide emissions, and depletion of the ozone shield. All three problems, unless seriously addressed by citizens, local, state, and Federal governments, will have permanent negative effects on the quality of life worldwide.

Acid precipitation alters the abiotic conditions of aquatic ecosystems so completely that lakes in the United States, Canada, and Europe are devoid of all life. Acid precipitation also alters nutrient cycling in terrestrial ecosystems leading to loss of plant life and erosion of topsoil. The sources of acid precipitation are well known. There is no doubt concerning cause-and-effect relationships! Slow or no responses by the government and industry have allowed continued and increased dumping of acid forming "oxides" into the atmosphere. It is easy, but regrettable, to make the predictions of expected outcomes if acid precipitation continues.

If you thought it was hot last summer, do not be surprised if global temperatures continue to rise and summer temperatures get hotter and hotter and hotter. Try to imagine the environmental and sociological impacts of melting polar ice caps, the bread basket of the world turning into a desert, and long-term droughts. These events have a high probability of occurring in your lifetime if reduced carbon dioxide emissions are not on the agendas of government and industry.

As if acid rain and the greenhouse effects were not enough to concern us, we must also contend with the potential of increased exposure to ultraviolet light - the ultimate suntan. The natural ecosystem provided a perfect shield (ozone) from ultraviolet light in the upper atmosphere. The human ecosystem has seen fit to destroy this life saving shield. Our actions regarding ozone are the ultimate paradox. We produce ozone at ground level as one component in photochemical pollution and destroy it in the place where it belongs and is of great benefit to us. The carcinogenic and mutagenic effects of ultraviolet light are well known. All life on Earth will be directly threatened by increasing ultraviolet radiation caused by depletion of the ozone shield.

There are solutions to the above three problems! In this chapter we will examine the causes of these three problems and what can be done to mitigate them.

STUDY QUESTIONS

ACID PRECIPITATION

1. Any precipitation that is more acid than normal is called _______________ **precipitation.**

2. Industrialized regions of the world are regularly experiencing precipitation that is from 10 to ___________ times more acid than normal.

3. List four forms of acid precipitation.

   a. ____________________, b. ____________________, c. ________________

   d. ____________________

UNDERSTANDING ACIDITY

Acids, Bases, and Water

4. Indicate whether the following are properties of [a] acids, [b] bases, or [c] water.

   [ ] Any chemical that will release OH^- ions.
   [ ] Any chemical that will release H^+ ions.
   [ ] The combination of one OH^- and 2 H^+ ions.
   [ ] The balance or neutral point between acidic and basic solutions.

pH: THE MEASUREMENT OF ACIDITY

5. **pH** is a measurement of (Hydrogen, Hydroxyl) ions in solution.

6. On the pH scale, the number __________ represents the pH of pure water or neutral.

7. On the pH scale, numbers decreasing from 7 (e.g., 6, 5, 4, 3, 2, 1) represent increasing (Acidity, Alkalinity).

8. Each unit on the pH scale represents a factor of (1, 10, 100, 1000) in the acid concentration.

9. A solution of pH 5 is (1, 10, 100, 1000) times more acid than a solution of pH 6.

10. A solution of pH 4 is (1, 10, 100, 1000) times more acid than a solution of pH 6.

EXTENT AND POTENCY OF ACID PRECIPITATION

11. Acid precipitation is defined as any precipitation with a pH of __________ or less.

12. Give the pH values for the following examples.

   a. Snowfall over large portions of the Eastern U.S. __________
   b. Fog in the Los Angeles area __________
   c. The lowest recorded pH in industrialized regions __________

THE SOURCE OF ACID PRECIPITATION

13. About two thirds of the acid in acid precipitation is ____________________ acid and one third is ____________________ acid.

14. The presence of these two acids should lead one to conclude that the source of the problem must be ____________________ and __________________ emissions into the air.

15. It is well known that sulfur and nitrogen are found in fossil ________________.

16. Heavy rains are (More, Less) acidic than fogs and mists.

17. Acid dews may result from ________________ fallout.

18. The major users of fossil fuels are ________________, ________________, and ________________.

19. Since three fourths of the sulfur dioxide comes from coal-burning, over __________ percent of the acid precipitation falling over Eastern U.S. and Canada can be attributed to coal-burning ________________ plants.

Need we go further on cause-and-effect relationships?

20. These power plants attempted to alleviate sulfur dioxide emission at ground level by building (Shorter, Taller) stacks that ultimately dispersed the pollutant (Closer, Further) from the source.

HOW ACID PRECIPITATION AFFECTS ECOSYSTEMS

Effects on Aquatic Ecosystems

21. pH controls all aspects of ________________, ________________, and ________________.

22. Indicate whether the following conditions are [+] or are not [-] known effects of acid pH in aquatic ecosystems.

[ ] Alteration of plant and animal reproduction.
[ ] Introduction of other toxic elements (e.g. aluminum) from soil leachate.
[ ] Shift from eutrophic to oligotrophic conditions.
[ ] Totally barren and lifeless aquatic ecosystems.
[ ] Alterations in the food chains.

Effects on Forests

23. Indicate whether the following effects of acid precipitation on forests are [a] direct contact, [b] removing nutrients, or [c] mobilization of toxic elements.

[ ] Increased leaching of nutrients.
[ ] The break down and release of aluminum into solution.
[ ] Retarded decomposition and nitrogen fixation.
[ ] Reduced growth and diebacks of plant and animal populations.
[ ] Increased vulnerability to natural enemies and drought.
[ ] Increased water loss through leaves.

24. There is probably a synergistic interaction between ________________ and acid precipitation.

DEPLETION OF BUFFERING CAPACITY AND ANTICIPATION OF FUTURE IMPACTS

25. A substance that absorbs hydrogen ions at a given pH is called a ________________.

26. Hydrogen ion concentrations in solution will (Increase, Decrease) in the presence of a buffer.

27. A mineral that is an important buffer in nature is ________________, known chemically as calcium carbonate.

28. The amount of acid that a given amount of limestone can neutralize is known as its _______________ capacity.

29. When an ecosystem loses its buffering capacity

a. additional hydrogen ions will remain in solution. (True, False)
b. the buffer is used up. (True, False)
c. the ecosystem will become acidified. (True, False)

30. When the buffering capacity is exhausted, there is a (Rapid, Slow) drop in pH.

EFFECTS ON HUMANS AND THEIR ARTIFACTS

31. Indicate whether the following effects of acid precipitation on humans is related to [a] artifacts, [b] health, or [c] aesthetics.

[ ] mobilization of aluminum and other toxic elements
[ ] corrosion of limestone and marble
[ ] deterioration of lakes and forests

STRATEGIES FOR COPING WITH ACID PRECIPITATION

Indicate whether the following strategies for coping with acid precipitation are examples of

[a] treating the symptoms
[b] reducing acid-forming emissions
[c] prescriptions for biosphere collapse
[d] what you can do

32. [ ] fuel switching

33. [ ] Demand immediate legislation to control sulfur dioxide emissions.

34. [ ] Failure to provide definitive proof of cause-and-effect relationships.

35. [ ] Costs of environmental destruction if emissions are not controlled.

36. [ ] scrubbers

37. [ ] alternative power plants

38. [ ] Adding lime to the soil.

Indicate whether the following strategies for reducing acid-producing emissions are examples of [a] fuel switching, [b] coal washing, [c] fluidized bed combustion, [d] scrubbers, or [e] alternative power plants.

39. [ ] Exhausting fumes through a spray of water containing lime.

40. [ ] Combustion of coal in a mixture of sand and lime.

41. [ ] Building more nuclear power plants.

42. [ ] Pulverize and chemically wash coal.

43. [ ] Using low-sulfur coals.

44. Utility companies have (More, Less) governmental support than does the public on the acid precipitation issue.

THE CARBON DIOXIDE GREENHOUSE EFFECT

45. Release of carbon dioxide into the atmosphere will have a gradual (Warming, Cooling) effect on global temperatures.

THE HEAT TRAPPING EFFECT OF CARBON DIOXIDE

46. Light energy is (Blocked, Absorbed) by the Earth's atmosphere.

47. Light energy is converted to ________________ energy in the form of infrared radiation.

48. When heat energy is trapped in an enclosed space and the temperature rises, this is known as the ____________________ effect.

49. Carbon dioxide (Repels, Absorbs) infrared radiation.

50. As carbon dioxide absorbs the infrared radiation it (Warms, Cools) the rest of the atmosphere.

51. The carbon dioxide concentrations in the Earth's atmosphere are (Increasing, Decreasing).

52. As carbon dioxide concentrations increase, global temperatures will (Increase, Decrease).

SOURCES OF CARBON DIOXIDE ADDITIONS

53. Photosynthesis (Adds, Subtracts) carbon dioxide from the atmosphere.

54. Respiration and burning of organic matter (Adds, Subtracts) carbon dioxide from the atmosphere.

55. Every pound of fossil fuel that is burned results in the production of an additional __________ pounds of carbon dioxide.

56. A total of __________ billion tons of carbon dioxide is added annually to the Earth's atmosphere from burning fossil fuels and tropical rain forests.

DEGREE OF WARMING AND ITS PROBABLE EFFECTS

57. Best estimates of the overall warming of the Earth's atmosphere in the next 50 years will be between __________ and __________ degrees centigrade.

58. Indicate whether the following are [+] or are not [-] probable effects of global warming.

    [ ] Melting of polar ice caps.
    [ ] Flooding of coastal areas.
    [ ] Massive migrations of people inland.
    [ ] Alteration of rainfall patterns.
    [ ] Deserts becoming farmland and farmland becoming deserts.
    [ ] Significant losses in crop yields.

IS THE CARBON DIOXIDE WARMING EFFECT HERE?

59. Is it (True, False) that five of the hottest years on record have occurred in the last decade?

60. Is it (True, False) that carbon dioxide levels in the atmosphere are rising?

STRATEGIES FOR COPING WITH THE CARBON DIOXIDE EFFECT/WHAT YOU CAN DO

61. Which of the following are strategies for coping with the carbon dioxide effect?

a. increase fuel efficiency, b. use non-fossil fuel energy, c. halt deforestation
d. plant trees, e. all of these are strategies

DEPLETION OF THE OZONE SHIELD

THE NATURE AND IMPORTANCE OF THE OZONE SHIELD

62. Ultraviolet light is responsible for thousands of cases of ________________ cancer per year in the U.S.

63. ________________ in the stratosphere prevents ultraviolet light from entering the Earth's atmosphere.

64. The presence of this chemical barrier to ultraviolet light is known as the ________________ shield.

65. Is it (True, False) that the ozone in the stratosphere and the ozone that is a serious pollutant on the Earth's surface are the same ozone?

THE FORMATION AND BREAKDOWN OF THE OZONE SHIELD

66. Complete the following reactions that form ozone.

Reaction #1: ________________ light + O^2 ----> free ________________ atoms

Reaction #2: free ________________ atoms + O^2 ----> ________________ (O^3)

Reaction #3: free ________________ atoms + O^3 ----> ________________ atoms

67. A ________________ is a chemical which promotes a chemical reaction without itself being used up by the reaction.

68. Chlorine is a catalyst that destroys the product of which of the above (question #66) reactions?

a. Reaction #1, b. Reaction #2, c. Reaction #3

69. Less ozone in the stratosphere will cause (More, Less) UV light penetration.

THE SOURCE OF CHLORINE ATOMS ENTERING THE STRATOSPHERE

70. The source of chlorine entering the stratosphere is ________________.

71. Which of the following industries are sources of chlorofluorocarbons (CFCs)?

a. refrigeration, b. plastic, c. electronics, d. cosmetic, e. all of these industries are sources

72. Is it (True, False) that CFCs will eventually be flushed out of the atmosphere?

73. Is it (True, False) that CFCs are water insoluble?

THE OZONE "HOLE"

74. A gaping hole in the ozone shield was discovered over the ________________ pole.

75. Is it (True, False) that this hole persists throughout the year?

COMING TO GRIPS WITH OZONE DEPLETION

76. Is it (True, False) that several nations and industries are scaling back CFCs production?

77. List two actions you can take to reduce the rate of ozone depletion.

a. ____

b. ____

KEY WORDS AND CONCEPTS

acid rain
greenhouse effect
acid
absorption
hydroxyl ion
ultraviolet radiation
pH scale
acid pH
neutral pH
mobilization
nitrogen oxides
fluidized bed combustion
carbon dioxide effect
acid deposition
light radiation
fossil fuels
ozone
pH
chlorine
basic pH
buffer
acid leaching of nutrients
coal washing
acid precipitation
infrared radiation
hydrogen ion
base
neutralization
ozone shield
catalyst
CFCs
buffering capacity
sulfur dioxide
scrubbers

SELF TEST

Circle the correct answer to each question.

1. Any chemical that releases H^+ ions is

a. an acid, b. a base, c. water, d. all of these

2. Measurement of pH is actually a measure of ________ in solution.

a. hydrogen ions, b. hydroxyl ions

3. A solution of pH 5 is ____________ times more acid than a solution of pH 6.

   a. 1, b. 10, c. 100, d. 1000

4. About two thirds of the acid in acid precipitation is

   a. chromic acid, b. nitric acid, c. sulfuric acid, d. carboxylic acid

5. It is well known that the source of acid precipitation is

   a. sulfur dioxide in coal.
   b. nitrogen oxides in gasoline.
   c. coal fired power plants.
   d. all of these

6. The effects of acid precipitation on aquatic ecosystems does not include

   a. alteration of plant and animal reproduction.
   b. introduction of toxic elements such as aluminum.
   c. a shift from eutrophic to oligotrophic conditions.
   d. the production of a totally barren and lifeless aquatic ecosystem.

7. Increased leaching of nutrients is an effect of acid precipitation on forests known as

   a. direct contact of acid rain   b. nutrient removal
   c. mobilization of toxic elements   d. all of these

8. The break down and release of aluminum into solution is an effect of acid precipitation on forests known as

   a. direct contact of acid rain   b. nutrient removal
   c. mobilization of toxic elements   d. all of these

9. Switching to alternative fuels is a strategy for coping with acid precipitation known as

   a. treating the symptoms   b. reducing acid-forming emissions
   c. prescriptions for biosphere collapse   d. what we can do

10. Exhausting fumes through a spray of water containing lime is a strategy for coping with acid precipitation known as

    a. treating the symptoms   b. reducing acid-forming emissions
    c. prescriptions for biosphere collapse   d. what we can do

11. There is ample evidence to suggest a direct correlation between global warming and increased

    a. carbon dioxide concentrations, b. ozone concentrations, c. acid rain, d. CFC's

12. Which of the following is not an expected effect of global warming?

    a. Melting of polar ice caps.
    b. Flooding of coastal areas.
    c. Increased crop yields.
    d. Alteration of rainfall patterns.

13. The ozone shield in the stratosphere protects Earth's biota from the damaging effects of

    a. CFC's, b. UV light, c. global warming, d. acid precipitation

14. The loss of the ozone shield can be attributed to the presence of _________ in the stratosphere.

    a. CFC's, b. oxygen, c. acid, d. carbon dioxide

15. The industrial source of chlorofluorocarbons in the stratosphere is

    a. refrigeration, b. plastic, c. cosmetic, d. all of these

---

ANSWERS TO STUDY QUESTIONS

1. acid; 2. 1000; 3. rain, fog, mist, snow; 4. b, a, c, c; 5. hydrogen; 6. 7; 7. acidity; 8. 10; 9. 10; 10. 100; 11. 5.5; 12. 4.5, 2.8, 1.5; 13. sulfuric, nitric; 14. sulfur dioxide, nitrogen oxides; 15. fuels; 16. less; 17. acid; 18. power plants, vehicles, industry; 19. 50, power; 20. taller, further; 21. enzymes, hormones, proteins; 22. +, +, -, +, +; 23. b, c, b, d, d, a; 24. ozone; 25. buffer; 26. decrease; 27. limestone; 28. buffering; 29. all true; 30. rapid; 31. b, a, c; 32. b; 33. d; 34. c; 35. c; 36. b; 37. b; 38. a; 39. d; 40. c; 41. e; 42. b; 43. a; 44. more; 45. warming; 46. absorbed; 47. heat; 48. greenhouse; 49. absorbs; 50. warms; 51. increasing; 52. increase; 53. subtracts; 54. adds; 55. 3; 56. 7.6; 57. 1.5, 4.5; 58. all +; 59. true; 60. true; 61. skin; 63. ozone; 64. ozone; 65. true; 66. (UV, oxygen), (oxygen, ozone), (oxygen, oxygen); 67. catalyst; 68. b; 69. more; 70. CFC's; 71. e; 72. false; 73. true; 74. south; 75. false; 76. true; 77. demand recapture of freon, boycott use of foam products

---

ANSWERS TO SELF TEST

1. a; 2. a; 3. b; 4. c; 5. d; 6. c; 7. b; 8. c; 9. b; 10. b; 11. a; 12. c; 13. b; 14. a; 15. d

CHAPTER 14

RISKS AND ECONOMICS OF POLLUTION

At this point in your study of environmental science you should be developing an increasing awareness of the serious pollution problems that the human ecosystem is imposing on natural ecosystems. Most water and air pollution problems have direct linkage with the economies and cultural priorities within developed countries. For example, enormous profits are realized by industries that produce the **things** demanded by the consumer public. People in developed countries have a profit mentality! Chapter 14 demonstrates how this profit mentality can serve to reduce rather than enhance water and air pollution. You are aware of several technologies that could be used to control water and air pollution but numerous competing economic interests often block action. This is the case because industries may view the implementation of these technologies as unnecessary overhead costs that would reduce their profit margin. In fact, it is good business to control pollution. Imagine the amount of money lost to industries and stock holders when employees are too sick to work, health insurance costs rise, equipment deteriorates, and when the costs of raw materials rise because of limited supplies. What are the broad, long-term risks associated with continued water and air pollution and economic benefits that could be derived from the utilization of pollution abatement technologies? The objective of Chapter 14 is to examine this question.

STUDY QUESTIONS

THE COST-BENEFIT ANALYSIS

1. A comparison of costs and benefits is commonly called a _______________ ratio.

2. Cost-effective projects have benefits that are (Greater, Less) than the costs.

3. Is it (True, False) that most forms of pollution control involve additional expenses?

4. List two ways that pollution control increases costs.

   a. ____

   b. ____

5. List two benefits derived from pollution control (see Table 14-1).

   a. ____

   b. ____

6. Is it (True, False) that too much pollution cleanup will outweigh benefits?

7. It is only when pollutant levels reach the ___________________ level that harmful effects are noticed and increase rapidly.

8. Optimum cost-effectiveness is achieved when the benefit curve is the (Greatest, Least) distance above the cost curve (see Figure 14-1c).

PROBLEMS IN PERFORMING COST-BENEFIT ANALYSIS

ESTIMATING COSTS

9. Is it (True, False) that equipment, labor, and maintenance costs can be estimated with a fair degree of objectivity?

10. Cost of pollution controls will be (High, Low) at the time they are initiated and tend to (Increase, Decrease) as time passes.

ESTIMATING BENEFITS AND PERFORMING RISK ANALYSIS

11. Indicate whether the following benefits or risks can [+] or cannot [-] be estimated with real dollar values.

    [ ] Health benefits from eliminating air pollution episodes.
    [ ] Costs of maintaining and replacing materials.
    [ ] Income from water recreation.
    [ ] Future risks of environmental degradation.

12. Risk analysis includes factors such as

    a. the number of people that may be affected. (True, False)
    b. the geographic extent of damage. (True, False)
    c. the nature and/or severity of the effects. (True, False)
    d. the probability of events occurring. (True, False)
    e. immediacy of the threat. (True, False)
    f. indirect effects of the threat. (True, False)
    g. potential reversibility of the threat. (True, False)

13. The *value* obtained in risk assessments is the ____________________ between the cost of damage that might occur in the absence of controls and mitigated costs occurring with controls.

PROBLEMS IN COMPARING COSTS AND BENEFITS

14. Is it (True, False) that some situations that may appear to be cost-ineffective in the short term may be extremely cost-effective in the long term?

15. Identify two types of pollution that may apply to a "true" response to question #14.

    a. ____

    b. ____

16. Is it (True, False) that those who bear the cost of pollution control and those who benefit from these controls are different groups of people?

17. Give one example of a situation that would support a "true" response to question #16.

    a. ____

THE NEED FOR REGULATION AND ENFORCEMENT

18. Is (Conscience, Economic Pressures) the most likely factor to influence an industry's decision to control pollutants?

19. An answer of "economic pressures" to question #18 has made it necessary to institute ________________ and ___________________ to affect all offending companies equally.

CONCLUSION

20. Ultimately all of __________________ will receive the benefits of environmental protection and suffer the costs of environmental degradation.

KEY VOCABULARY AND CONCEPTS

benefit-cost ratio	cost-effective	threshold level
cost-benefit analysis	estimating costs	risk analysis
short-term view	long-term view	regulation
enforcement		

SELF TEST

Circle the correct answer to each question.

1. An attempt to evaluate the benefits and drawbacks of a given project or course of action is referred to as

   a. an argument, b. a cost-benefit analysis, c. benefit-cost ratio, d. indecisiveness

2. Increased costs of pollution control would probably not include

   a. purchase price of new equipment.
   b. retraining personnel to use pollution control equipment.
   c. losses incurred from obsolete equipment.
   d. increased health care costs.

3. Which of the following would not be a direct benefit derived from pollution control?

   a. Improved public health.
   b. Reduced loss of material goods through corrosion and deterioration.
   c. Decreased real estate values.
   d. Increased recreational use of natural areas.

4. Which of the following is the best expression of the relationship between benefits and pollution control?

   a. There are no benefits until 100% control is achieved.
   b. Benefits are significant with modest control measures but tend to diminish as pollution control reaches 100%.
   c. There is a one-to-one relationship between benefits and control measures.
   d. The benefit curve is exponential to the arithmetic curve of control costs.

5. The optimum cost-effectiveness of costs for pollution control is generally achieved at

a. near-zero level of pollution control.
b. a near 100% level of pollution control.
c. a level somewhere between a and b above.
d. a 100% pollution control level.

6. The threshold level of pollutant refers to

a. a level where no harmful effects are observed.
b. a level that demands caution because danger is imminent.
c. the level where harmful effects are noticed and increase rapidly.
d. a level used to judge the standard index of a pollutant.

7. Estimating potential damage if pollution is not controlled is known as

a. cost-benefit analysis.
b. risk analysis.
c. the threshold level.
d. a determination of cost-effectiveness.

8. The benefits derived from pollution control

a. are greatest in the short-term and diminish in the long-term.
b. are low in the short-term but increase exponentially the longer the control is in effect.
c. are minimal at first, then increase to a maximum, then decrease again.
d. are maximal at first, then decrease to a minimum, then increase again.

9. Risk analysis does not directly include assessments of

a. the number of people that may be affected.
b. the geographic extent of damage possible.
c. costs of maintaining and replacing materials.
d. indirect effects of a pollution threat.

10. A final resolution to which of the following pollution problems would be cost-ineffective in the short-term but extremely cost-effective in the long-term?

a. acid rain, b. ground water contamination, c. loss of ozone shield, d. all of these

---

ANSWERS TO STUDY QUESTIONS

1. benefit-cost; 2. greater; 3. true; 4. equipment purchase and loss of old equipment, implementing control strategies; 5. improved public health, reduced corrosion and deterioration of materials, reclaiming recreational use of polluted areas; 6. true; 7. threshold; 8. greatest; 9. false; 10. high, decrease; 11. all +; 12. all true; 13. difference; 14. true; 15. acid rain, ground water pollution; 16. true; 17. pollutant produced in one state has greatest effect in another; 18. economic; 19. laws, regulations; 20. society

---

ANSWERS TO SELF TEST

1. b; 2. d; 3. c; 4. b; 5. c; 6. c; 7. b; 8. a; 9. c; 10. d

CHAPTER 15

THE PESTICIDE TREADMILL

The struggle between man and insects began long before the dawn of civilization, has continued without cessation to the present time, and will continue, no doubt, as long as the human race endures. We commonly think of ourselves as the lords and conquerors of nature. But insects had thoroughly mastered the world and taken full possession of it before man began the attempt. They had, consequently, all the advantages of possession of the field when the contest began, and they have disputed every step of our invasion of their original domain so persistently and successfully that we can even yet scarcely flatter ourselves that we have gained any very important advantage over them. If they want our crops, they still help themselves to them. If they wish the blood of our domestic animals, they pump it out of the veins of our cattle and our horses at their leisure and under our very eyes. If they choose to take up their abode with us, we cannot wholly keep them out of the houses we live in. We cannot even protect our very persons from their annoying and pestiferous attacks, and since the world began, we have never yet exterminated---we probably shall never exterminate---so much as a single insect species.

(THE BUGS ARE COMING, Time Magazine, July 12, 1976, page 38.)

When someone says the word "pest" to you, what is the first image that comes to mind? For most people, the image is a bug! This chapter and the next will broaden you understanding of the sorts of organisms that are considered pests in the human ecosystem. In fact, you will see how the structure and function of the human ecosystem is predestined to be plagued by all sorts of pests. The very definition of a pest, i.e., "any organism that impacts negatively on human health or economics", should help you to visualize a whole host of organisms that would fit this definition. For example, organisms that cause disease, harass us, feed on our crops and domestic livestock, cause the deterioration of our goods and food, or detract from our quality of life in any way are **pests**. Pest organisms can be plants, animals, bacteria, and parasites among others.

It is an easy task to identify the pest. It is much more difficult to figure the best procedures for eradicating it or, more importantly, how to control a pest and maintain the integrity of the ecosystem. Chapter 15 provides a historical overview of our early attempts at pest control that did not take the integrity of the ecosystem into consideration. Chapter content demonstrates how quick and dirty solutions to a complex problem can back fire. It is important to emphasize that the early attempts at pest control were complete failures and left the pests in a more virulent condition. Additionally, the natural ecosystem was inundated with vast quantities and types of chemicals (pesticides) which had the most adverse effects on non-pest organisms.

STUDY QUESTIONS

PROMISES AND PROBLEMS OF CHEMICAL PESTICIDES

1. The term "cide" means (a. to heal, b. to hurt, c. to kill, d. to save, e. to eradicate).

2. Indicate whether the following chemicals are meant to target [a] all organisms, [b] pests in general, [c] mice and rats, [d] insects, or [e] fungi.

   [ ] fungicides, [ ] rodenticides, [ ] insecticides, [ ] biocides, [ ] pesticides

DEVELOPMENT OF CHEMICAL PESTICIDES AND THEIR APPARENT VIRTUES

3. One of the first chemical pesticides used over 1000 years before the time of Christ was

   a. mercury, b. arsenic, c. copper sulfate, d. sulfur, e. DDT

4. Indicate whether the following are characteristics of [a] first-generation, [b] second-generation, or [c] both types of pesticides.

   [ ] Consisted of toxic heavy metals.
   [ ] Are inorganic compounds.
   [ ] Are synthetic organic compounds.
   [ ] Are persistent.
   [ ] Promote pest resistance.
   [ ] Are broad spectrum in their effect.
   [ ] Examples are lead, arsenic, and mercury.
   [ ] An example is DDT (dichloro diphenyl trichloroethane)

PROBLEMS STEMMING FROM CHEMICAL PESTICIDE USE

5. List four problems associated with the use of synthetic organic pesticides.

   a. ____

   b. ____

   c. ____

   d. ____

Development of Pest Resistance

6. Chemical pesticides gradually lose their __________________.

7. Loss of effectiveness leads to two choices by growers, namely, use (More, Less) or use the (Same, New) pesticides.

8. The use of chemical pesticides on insects leads to selection for the (Sensitivity, Resistance) gene in the gene pool of the population.

9. Each generation of insects exposed to a chemical pesticide will have (More, Less) genes for resistance in the gene pool of the population.

10. Is it (True, False) that as a pest population becomes resistant to one pesticide it may, at the same time, become resistant to others to which it has not yet been exposed?

Resurgences and Secondary Pest Outbreaks

11. Match the following definitions with [a] resurgence or [b] secondary pest outbreak.

    [ ] Populations of insects that were previously of no concern, start multiplying creating new problems.
    [ ] The return of the pest at higher and more severe levels.

12. Pesticide treatments often have a (Less, Greater) effect on the natural enemies of pests.

13. Which of the following statements are [+] or are not [-] explanations of why pesticides have a greater impact on natural enemies than upon the target pest?

[ ] Herbivore species are intrinsically more resistant to the pesticides than are its predators.
[ ] Biomagnification through the food chain gives the predator a higher dose of the pesticide.
[ ] Predatory species may be starved out due to a temporary lack of prey.

14. The chemical approach fails because it operates contrary to basic _______________ principles.

Increasing Costs

15. Is it (True, False) that the costs of pest control often outweigh the value of the product?

Adverse Environmental and Human Health Effects

16. What were three sources of evidence that indicated DDT was the causative factor in reproductive failure in fish-eating birds?

a. DDT levels in fragile egg shells were (High, Low).
b. DDT (Promotes, Reduces) calcium metabolism.
c. DDT levels in organisms (e.g., fish-eating birds) is (Higher, Lower) at the top of the food chain.

17. DDT is a ________________ hydrocarbon.

18. DDT accumulates in the body ______________ of humans and other animals.

19. DDT was banned in the United States in the early 19_________s.

20. Is it (True, False) that when DDT was banned, pesticides were no longer used?

21. About _________ billion pounds of pesticides were used in the United States in 1982.

22. Less than _________ percent of the pesticides dumped on the environment come into contact with the pest organism.

23. Where does the remaining 99 percent of the pesticide go?

a. Drifts into the air? (True, False)
b. Settles on surrounding ecosystems? (True, False)
c. Settles on bodies of water? (True, False)
d. Remains as residue on foods? (True, False)
e. Leaches from the soil into aquifers? (True, False)
f. Is captured and recycled by growers? (True, False)

THE PESTICIDE TREADMILL

24. Construct the last four steps in the below pesticide treadmill beginning with [a] new and larger quantities of chemicals followed by [b] more resistance and secondary outbreaks.

[a] Step 1 -> [b] Step 2 -> [ ] Step 3 -> [ ] Step 4 -> [ ] Step 5 -> [ ] Step 6 -> ?

ATTEMPTING TO PROTECT HUMAN AND ENVIRONMENTAL HEALTH

THE FEDERAL INSECTICIDE, FUNGICIDE, AND RODENTICIDE ACT (FIFRA)

25. What is the law that regulates pesticide use? ______________________________

26. The law requires that manufacturers ____________ pesticides with the government before marketing them.

27. The registration procedure requires ____________ to determine toxicity to animals.

SHORTCOMINGS OF FIFRA

28. List four shortcomings of FIFRA.

    a. ____

    b. ____

    c. ____

    d. ____

29. Indicate whether the following statements are reflections of shortcomings in FIFRA related to [a] inadequate testing, [b] bans on a case-by-case basis, [c] pesticide exports, or [d] lack of public input.

    [ ] Kepone in the James River of Virginia.
    [ ] 1400 chemicals and 40,000 formulations.
    [ ] 100 million pounds of pesticides shipped to less developed countries.
    [ ] EDB found in hundreds of wells in Florida and California.
    [ ] Using 100 to 200 mice over a one- to two-year period.
    [ ] Substituting healthy mice for experimental mice that died.
    [ ] Intense lobbying efforts by the chemical industry as opposed to public forums.

NONPERSISTENT PESTICIDES: ARE THEY THE ANSWER?

30. Suppose the half-life of DDT is 20 years and that 1000 pounds was applied to a field in Nebraska in 1950. How much of the 1000 pounds will be left in:

    a. 1970 ____________, b. 1990 ____________, c. 2010 ____________

    d. Will the DDT ever completely disappear? (Yes, No)

31. The half-life of nonpersistent pesticides is (a. days, b. weeks, c. years, d. a or b).

32. Indicate whether the following statements are true [+] or false [-] concerning nonpersistent pesticides.

    [ ] Stay in the environment for a long time.
    [ ] Are less toxic than DDT.
    [ ] Are broad spectrum pesticides.
    [ ] Do not cause resistance in pest species.
    [ ] Do not cause resurgence or secondary outbreaks.

KEY VOCABULARY AND CONCEPTS

pest
first-generation pesticide
broad-spectrum
resurgence
broken bird eggs
inadequate testing
pesticide
second-generation pesticide
persistent
secondary outbreaks
pesticide treadmill
herbicides
biocide
DDT
pest resistance
biomagnification
FIFRA
nonpersistent pesticides

SELF TEST

1. An insecticide kills

   a. insects, b. all pests, c. other organisms, d. all of these

2. One of the earliest pesticides was

   a. mercury, b. arsenic, c. sulfur, d. DDT

3. First-generation pesticides

   a. consisted of inorganic compounds.
   b. were used after World War II.
   c. are synthetic organic compounds.
   d. are exemplified in DDT.

4. Second-generation pesticides

   a. consisted of inorganic compounds.
   b. were used prior to World War II.
   c. are synthetic organic compounds.
   d. are exemplified in arsenic.

5. The first synthetic organic pesticide used was

   a. DDT, b. kepone, c. mirex, d. chlordane

6. Apparent advantage(s) of DDT included that it was

   a. broad spectrum, b. persistent, c. inexpensive to produce, d. all of these

7. Which of the following is not a problem associated with the use of synthetic organic compounds?

   a. The return of the pest at higher and more severe levels.
   b. Populations of insects that were previously of no concern, start multiplying creating new problems.
   c. The need to continually add more pesticide because it does not stay around very long.
   d. One compound is effective against a wide variety of insects.

8. Which of the above answer choices (#7, a - d) is a definition of secondary outbreaks?

9. The broad-spectrum characteristic of pesticides means

a. the return of the pest at higher and more severe levels.
b. populations of insects that were previously of no concern, start multiplying creating new problems.
c. the need to continually add more pesticide because it does not stay around very long.
d. one compound is effective against a wide variety of insects.

10. The need to continually add more pesticide is not the result of

a. short half-lives of pesticides.
b. development of pest resistance.
c. loss of pesticide effectiveness.
d. decreased pest sensitivity to pesticides.

11. Which of the following has not been a result of using pesticides against insect pests?

a. Development of pest resistance.
b. Total eradication of the pest.
c. Decreased pest sensitivity to pesticides.
d. Loss of pesticide effectiveness.

12. Predatory insects are likely to be more severely affected by pesticide treatments than the herbivore insects because

a. herbivore insects are intrinsically more resistant to the pesticides than are its predators.
b. predatory insects receive a larger dose of the pesticide through the food chain.
c. predatory insects may be starved out due to a temporary lack of prey.
d. All of the above are reasons.

13. Which of the following is not a characteristic of nonpersistent pesticides?

a. Stay in the environment for a long time.
b. Can be more toxic than DDT.
c. Are broad spectrum pesticides.
d. Cause the development of resistance in pest species.

14. Which of the following statements does not describe characteristics of the pesticide treadmill?

a. Using increasing amounts and types of new chemicals.
b. Increasing risks to human and environmental health.
c. Increasing control over selected pests.
d. Increasing resistance and secondary pest outbreaks.

15. Which of the following shortcomings in FIFRA promotes heavy lobbying by the chemical industry?

a. Inadequate testing.
b. Lack of public input.
c. Pesticide exports.
d. Ban issued on a case-by-case basis when threats are proven.

ANSWERS TO STUDY QUESTIONS

1. c; 2. e, c, d, a, b; 3. d; 4. a, a, b, c, c, c, a, b; 5. pest resistance, resurgence, cost, adverse human health and environmental effects; 6. effectiveness; 7. more, new; 8. resistance; 9. more; 10. true; 11. b, a; 12. greater; 13. all +; 14. ecological; 15. true; 16. high, reduces, higher; 17. halogenated; 18. fat; 19. 70; 20. false; 21. 1.7; 22. one; 23. (a - e, all true), f = false; 24. a, b, a, b; 25. FIFRA; 26. register; 27. tests; 28. inadequate testing, proof of cause-and-effect, pesticide exports, lack of public input; 29. b, a, c, b, a, a, d; 30. 500, 250, 125; 31. d; 32. -, -, +, -, -

ANSWERS TO SELF TEST

1. d; 2. c; 3. a; 4. c; 5. a; 6. d; 7. c; 8. b; 9. d; 10. a; 11. b; 12. d; 13. a; 14. c; 15. b

CHAPTER 16

NATURAL PEST CONTROL METHODS AND INTEGRATED PEST MANAGEMENT

Your study of the last chapter must have left you with some feelings of how utterly futile our past efforts of pest control have been. Chemical control methods have certainly won lots of battles with pests but we are still losing the war! We need to develop an entirely different arsenal and strategy to effectively **manage** pest species. Note that the emphasis is on management not eradication. We will probably never eradicate a single insect pest. Chapter 16 addresses both the arsenal and strategy options leading to natural or biological rather than chemical controls. Overall, the emphasis is on ecosystem stability - a concept you have studied extensively in previous chapters.

STUDY QUESTIONS

1. Ecological pest management concentrates on the manipulation of ________________ factors instead of the use of synthetic chemicals.

2. Management techniques that work with natural factors are called ____________________ or ____________________ control.

THE INSECT LIFE CYCLE AND ITS VULNERABLE POINTS

3. A first step in natural or biological control is to understand the ____________________ cycle of pest species.

4. Affecting any stage in the life cycle can prevent __________________ and affect control.

METHODS OF NATURAL OR BIOLOGICAL CONTROL

5. List the five ways of natural or biological pest control

   a. ____

   b. ____

   c. ____

   d. ____

   e. ____

CONTROL BY NATURAL ENEMIES

6. Match the following examples of pests with the natural enemy that was used to control each pest.

Pest	Natural Enemy
[ ] scale insects	a. manatees
[ ] caterpillars	b. plant-eating insects
[ ] gypsy moths and Japanese beetles	c. bacteria
[ ] prickly pear cactus	d. parasitic wasps
[ ] water hyacinths	e. vedalia (ladybird) beetles
[ ] rabbits in Australia	

7. The first step in using natural enemies is to ______________ the natural enemies that already exist.

8. Of the 2000 serious insect pests, _________ percent remain without effective natural enemies.

9. Is it (True, False) that finding natural enemies to pests generates as much profit for industry as do synthetic chemicals?

10. Natural enemies are a (Short, Long) term form of pest control.

11. Natural enemies are (More, Less) profitable in the long term than synthetic chemicals.

GENETIC CONTROL

12. Prey species react the (Same, Differently) to attacks by predators.

13. Prey susceptibility to predators is the (Same, Different).

14. There is a genetic (Compatibility, Incompatibility) between the pest and species that are not attacked.

15. The potential of a form of ________________ control is implied by genetic incompatibility between the pest and species that are not attacked.

Chemical Barriers

16. A chemical barrier means that the plant produces some chemical that is ________________ or _______________ to potential pests.

17. A chemical barrier was introduced into wheat plants that caused Hessian fly ______________ to die as they fed on leaves.

18. Is it (True, False) that some natural chemical barriers are highly toxic to humans?

Physical Barriers

19. Physical barriers are _________________ traits that impede pest attacks.

20. An example of a physical barrier is ________________ hairs on leaf surfaces.

21. Is it (True, False) that pests cannot overcome genetic controls through resistance?

THE STERILE MALE TECHNIQUE

22. The sterilization procedure involves subjecting the

a. egg, b. larvae, c. pupa, d. adults

to just enough high energy radiation to render them sterile.

23. Is it (True, False) that pests cannot develop resistance to the sterile male technique?

24. An example of application of the sterile male technique is the _______________ fly.

CULTURAL CONTROLS

25. Non-chemical alteration of one or more environmental factors producing a "foreign" environment to pests is called _______________ control.

Cultural Control of Human Pests

26. Indicate whether the following control measures are in the category of [a] sanitation or [b] personal hygiene.

[ ] water purification
[ ] bathing
[ ] wearing clean clothing
[ ] proper and systematic disposal of garbage
[ ] using clean cooking and eating utensils
[ ] refrigeration, freezing, and canning foods

27. It is estimated that _________ percent of human illness in less developed countries results from drinking contaminated water.

Cultural Control of Pests Affecting Lawns, Gardens, and Crops

Indicate whether the following control measures are examples of

[a] Selection of what to grow and where to grow it
[b] Management of lawns and pastures
[c] Water and fertilizer management
[d] Time of planting
[e] Destruction of crop residues
[f] Adjacent crops and weeds
[g] Crop rotation
[h] Polyculture
[i] Customs and quarantines

28. [ ] Prohibiting biological materials that carry pests from entering a country.

29. [ ] Planting plants under their optimum conditions.

30. [ ] Growing two or more species together or in alternate rows.

31. [ ] Cutting grass no less than 3 inches.

32. [ ] Alternate crops from one year to the next in a given field.

33. [ ] Providing the optimum (neither too much nor too little) levels of water and fertilizer.

34. [ ] Plowing under or burning the residue left after harvest.

35. [ ] Eliminate plants that act as pest attractants and grow others that act as repellents.

36. [ ] Delay planting so most of the pest population starves before the plants are available.

NATURAL CHEMICAL CONTROL

37. Indicate whether the following definitions pertain to [a] hormones or [b] **pheromones.**

[ ] A chemical secreted by one insect that influences the behavior of another.
[ ] Chemicals that provide signals for development and metabolic functions.

38. The four aims of natural chemical control are to ____________________, ____________________, ____________________, and ____________________ an insect's own hormones or pheromones to disrupt its life cycle.

39. The two advantages of natural chemicals are that they are

a. ____________________ and b. ____________________ .

40. Juvenile hormone had a direct adverse affect on the

a. egg, b. larva, c. pupa, d. adult

41. Sex attractant pheromones may be used in a ____________________ or ____________________ technique.

42. Is it (True, False) that the potential of future control of pests through natural chemicals is promising?

43. What regulatory hurdle now represents the greatest barrier to deployment of non-toxic compounds?

a. Clean Water Act, b. Clean Air Act, c. FIFRA, d. Conservation Reserve Act

ECONOMIC CONTROL VS. TOTAL ERADICATION

44. Natural controls are aimed at (Management, Eradication) of pest populations.

45. The economic threshold is a measure of ____________________ damage.

46. Significant damage is implied when the cost of damage is considerably (Greater, Less) than the cost of pesticide application.

CONCLUSION

47. Is it (True, False) that there is tremendous potential for controlling pest problems through natural means?

48. Is it (True, False) that there is considerable support for the development and deployment of natural controls?

49. Is it (True, False) that we continue on the pesticide treadmill?

INTEGRATED PEST MANAGEMENT

50. Integrated pest management (IPM) addresses all ____________________, ____________________, and ____________________ factors.

51. IPM is an/a (Approach, Technique).

SOCIOECONOMIC FACTORS SUPPORTING OVERUSE OF PESTICIDES

Indicate whether the following statements are examples of

[a] "The only good bug is a dead bug" attitude.
[b] Aesthetic quality and cosmetic spraying.
[c] Insurance spraying.
[d] Chemical company profits.

52. [ ] Consumer desire for unblemished produce.

53. [ ] The use of pesticides "just to be safe".

54. [ ] The satisfaction of watching insects drop dead.

55. [ ] Ignorance of whether a "bug" is causing damage or beneficial.

56. [ ] The "Snow White" syndrome.

57. [ ] Promoting and exploiting the above attitudes.

58. [ ] Continued resurgence, secondary pest outbreaks, and increasing pest resistance.

ADDRESSING SOCIOECONOMIC ISSUES

Indicate whether the following statements are examples of

[a] Produce labelling versus spraying
[b] Advice from trained field scouts and pest loss insurance
[c] Economics of natural controls

59. [ ] Identical crop production per acre on "organic" when compared to "conventional" farms.

60. [ ] Monitoring pest populations to determine whether they are exceeding the economic threshold.

61. [ ] Informed consumers abandon the Snow White attitude.

62. [ ] Food outlets selling "organically grown" produce.

INTEGRATION OF NATURAL CONTROLS

63. Indicate in what order (1 to 4) the following control measures are used to control Medflies.

    [ ] Spraying with Malathion.
    [ ] Using the sterile male technique.
    [ ] Using a sex attractant.
    [ ] Checking imported produce.

64. An accidental introduction of 100,000 additional pairs of fertile Medflies into California was the result of improper use of

    a. spraying with Malathion. b. using the sterile male technique. c. using a sex attractant.
    d. checking imported produce.

65. In the California infestation, public protests to the use of Malathion resulted in the use of (More, Less) of the chemical.

PRUDENT USE OF PESTICIDES

66. Is it (True, False) that the use of synthetic organic chemicals is not part of the IPM approach to pest management?

WHAT YOU CAN DO TO PROMOTE ECOLOGICAL PEST MANAGEMENT

67. When you have a pest problem in your own home or garden you should

    a. automatically use a synthetic chemical pesticide. (True, False)
    b. get advise from a pesticide dealer. (True, False)
    c. seek advise from an agricultural extension service agent. (True, False)
    d. seek to use natural controls first. (True, False)

KEY VOCABULARY AND CONCEPTS

insect life cycle
control by natural enemies
cultural control
physical barriers
economic factors
personal hygiene
customs
pheromones
trapping technique
economic threshold
Snow White attitude
field scouts
Medfly

natural control
genetic control
use of natural chemicals
structural traits
social factors
crop rotation
quarantines
juvenile hormone
confusion technique
IPM
vested interests
pest-loss insurance
Malathion

biological control
sterile male technique
chemical barriers
cosmetic spraying
sanitation
polyculture
hormones
sex attractant
economic control
insurance spraying
labelling
organic farming
FIFRA

SELF TEST

Circle the correct answer to each question.

1. Ecological pest management seeks to control pests by

   a. manipulating natural factors affecting the host and pest relationship.
   b. letting nature take its course.
   c. using pesticides that are known to be ecologically sound.
   d. using biological controls only.

2. The first step in natural or biological control is to understand

   a. safe versus unsafe pesticides.
   b. what the host and pest relationship is.
   c. the life cycles of pest species.
   d. how to prevent cosmetic spraying.

3. To prevent a pest insect from doing significant harm, it is sufficient to control

   a. metamorphosis, b. reproduction, c. pupation, d. adult females

   Indicate whether the following control strategies (#'s 4 to 8) are

   a. control by natural enemies, b. genetic controls, c. sterile male technique
   d. cultural control, or e. use of natural chemicals

4. Burning crop residue to destroy over wintering pests.

5. Developing a strain of wheat that causes pest larvae to die as they feed on the leaves.

6. Using a predatory insect on an insect pest.

7. Irradiating pupae larvae.

8. Baiting traps with pheromone from the female of the insect pest.

9. What is the first step in using natural enemies that already exist?

   a. identification, b. preservation, c. capture, d. captive breeding

10. A large percentage of the most serious pest species are those which

    a. have been introduced from other continents.
    b. are products of biotechnology.
    c. are recent mutations of non-pest species.
    d. have not been totally studied by entomologists.

11. Natural enemies of pest species are most likely found

    a. where the pest poses the most serious problem.
    b. where the pest species exists but is not a problem.
    c. in the tropics.
    d. in the farmbelt of the midwestern United States.

Match the control technique (a - e) with the pest organism that was controlled as a result of the use of the technique.

a. control by natural enemies, b. genetic controls, c. sterile male technique,
d. cultural control, or e. use of natural chemicals

12. Prickly pear cactus

13. Hessian fly larvae

14. Screw worm larvae

15. Water hyacinths

16. Crop rotation is a type of cultural control of pests that involves

a. planting crops under their optimum conditions.
b. growing two or more species together or in alternate rows.
c. alternating crops from one year to the next in a given field.
d. plowing under or burning the residue left after harvest.

17. Which of the following are part of the integrated pest management approach?

a. sociological factors, b. economic factors, c. ecological factors, d. all of these

18. Cosmetic spraying is done to

a. improve the appearance of produce.
b. improve the taste.
c. improve the storability.
d. protect the public health.

19. Which of the following measures to control medflies is the last one used?

a. Spraying with malathion.
b. Using the sterile male technique.
c. Using a sex attractant.
d. Checking imported produce.

20. If you have a pest problem in your own home or garden you should first

a. use a synthetic chemical pesticide.
b. seek advise from a pesticide dealer.
c. seek advise from an agricultural extension service agent.
d. use a nonpersistent organic pesticide.

ANSWERS TO STUDY QUESTIONS

1. natural; 2. natural, biological; 3. life; 4. reproduction; 5. control by natural enemies, genetic control, sterile male technique, cultural control, use of natural chemicals; 6. e, d, c, b, a, c; 7. preserve; 8. 90; 9. false; 10. long; 11. less; 12. differently; 13. different; 14. incompatibility; 15. genetic; 16. lethal, repulsive; 17. larvae; 18. true; 19. structural; 20. hooked; 21. true; 22. c; 23. true; 24. screw worm; 25. cultural; 26. a, b, b, a, a, a; 27. 80; 28. i; 29. a; 30. h; 31. b; 32. g; 33. c; 34. e; 35. f; 36. d; 37. b, a; 38. isolate, identify, synthesize, use; 39. species specific, non-toxic; 40. c; 41. trapping, confusion; 42. true; 43. c; 44. management; 45. significant; 46. greater; 47. true; 48. false; 49. true; 50. sociological, economic, ecological; 51. approach; 52. b; 53. c; 54. a; 55. c; 56. b; 57. d; 58. d; 59. c; 60. b; 61. a; 62. a; 63. 4, 3, 2, 1; 64. b; 65. more; 66. false; 67. false, false, true, true

ANSWERS TO SELF TEST

1. a; 2. c; 3. b; 4. d; 5. b; 6. a; 7. c; 8. e; 9. b; 10. a; 11. b; 12. a; 13. b; 14. c; 15. a; 16. c; 17. d; 18. a; 19. a; 20. c

CHAPTER 17

CONSERVATION OF NATURAL BIOTA

How many rivets do you think there are in a Boeing 747? There are probably thousands but each single rivet is as important as the thousands in terms of maintaining the structural integrity of the entire airplane. Now imagine the following series of events take place on your next trip. You have just boarded the plane and are comfortably settled into your seat. You look out the window and see a person on the wing. This person is taking one of the rivets out of the wing! You are (or should be) alarmed and call the stewardess. You complain that it is not a good thing for the person to be taking a rivet out of the wing of the plane. The stewardess calms your fears by telling you that you should not worry because there are thousands of rivets in the wing of this plane. She further tells you that the person extracting the rivets is getting $5.00 per rivet and will be extracting one rivet from the other wing also. Well, you take off and arrive safely at your destination. On your return flight, the same thing happens but now the person is extracting two rivets from each wing. The stewardess tells you that again it is alright because the person is now getting $10.00 per rivet. Well this story could go on and on but let's get to the point. How many rivets, regardless of the economic value, would you allow to be taken out of the plane before you fear for its structural integrity?

Let us ask the question in a manner that is more relevant to the conceptual framework of Chapter 17. How many species (rivets) will you allow to be eliminated from an ecosystem (Boeing 747) before you fear for its structural integrity? Previous chapters have stressed the need to maintain ecosystem stability. Each species of plant and animal plays a vital role in this process even though we may have not yet identified the importance of every single species. Nevertheless, we are rapidly pulling out the rivets that hold the ecosystem together through overharvest, destruction of habitats, and major alterations in the abiotic characteristics of ecosystems through pollution. Chapter 17 examines the values of natural biota, how the human ecosystem is sacrificing these values, and what needs to be done to preserve natural biota.

STUDY QUESTIONS

1. All species are collectively referred to as ________________.

2. We presently know of the existence of ________ million species and estimate that at least _________ to _________ million additional species have not been studied.

NATURAL BIOTA: TO PRESERVE OR DESTROY?

3. The manner and degree to which humans have used natural biota has changed through the ages. Match the following uses with the historical time frames of

   [a] primitive humans, [b] 10,000 years ago, [c] 6,000 years ago, [d] 250 years ago, [e] today

   [ ] Heat comes from a furnace and food from a grocery store.
   [ ] All food, clothing, fuel, and building material came from natural biota.
   [ ] Advent of agriculture.
   [ ] The beginning of the industrial revolution.
   [ ] Humans began learning to refine and fashion metals.

VALUES OF NATURAL BIOTA

4. Is it (True, False) that preserving natural biota will indicate a balanced biosphere?

5. List the five values of natural biota.

   a. ______________________________

   b. ______________________________

   c. ______________________________

   d. ______________________________

   e. ______________________________

Underpinnings of Agriculture

6. Indicate whether the following statements pertain to [a] wild populations of plants or animals or [b] cultivated (cultivar) populations of plants or animals.

   [ ] Highly adaptable to changing environmental conditions.
   [ ] Highly nonadaptable to changing environmental conditions.
   [ ] Have numerous traits for resistance to parasites.
   [ ] Lack genetic *vigor.*
   [ ] Can only survive under highly controlled environmental conditions.
   [ ] Have a high degree of genetic diversity in the gene pool of the population.
   [ ] Have a low degree of genetic diversity in the gene pool of the population.
   [ ] Represents a reservoir of genetic material commonly called a **genetic bank.**

Resource for Medicine

7. Of what value is the chemical called vincristine? ______________________________

   ____

8. In what plant is the chemical vincristine found? ______________________________.

9. Is it (True, False) that there is the potential of discovering innumerable drugs in natural plants, animals, and microbes?

Providing Natural Services

10. Indicate whether the following are [+] or are not [-] examples of natural services.

    [ ] Preventing erosion.
    [ ] Maintaining topsoil.
    [ ] Reducing runoff and flooding.
    [ ] Holding and recycling nutrients.
    [ ] Maintaining air quality.
    [ ] Moderating temperature.
    [ ] Provision of products of commercial and sport value.
    [ ] Pollination

11. Is it (True, False) that we can compensate for the losses of natural services?

Recreational, Aesthetic, and Scientific Values

12. Indicate whether the following are [a] recreational, [b] aesthetic, or [c] scientific values of the biota of natural ecosystems.

[ ] hunting
[ ] sport fishing
[ ] hiking
[ ] camping
[ ] Being in a forest rather than a city dump.
[ ] study of ecology
[ ] Just knowing that certain plant and animal species are alive.

Commercial Interests

13. As leisure time (Increases, Decreases), larger portions of the economy become connected to supporting activities related to the natural environment.

14. Pollution (Increases, Decreases) the commercial benefits of the natural environment.

15. Commercial interests often result in the (Conservation, Destruction) of natural biota.

ASSAULTS AGAINST NATURAL BIOTA

16. List four major assaults against natural biota.

a. ______________________________

b. ______________________________

c. ______________________________

d. ______________________________

Indicate whether the following examples represent

[a] loss of habitat, [b] pollution, [c] overuse, or [d] introduction of foreign species

types of assaults on natural biota.

17. [ ] Destruction of the American chestnut by chestnut blight disease.

18. [ ] Clearing of forests.

19. [ ] Overgrowth of forests by kudzu.

20. [ ] Draining or filling of wetlands for development or agriculture.

21. [ ] The carbon dioxide warming effect on global temperatures.

22. [ ] A growing population with growing demands for more land for housing and agriculture.

23. [ ] Prospects of huge immediate profits.

24. [ ] The growing fad for exotic pets, fish, reptiles, birds, and house plants.

25. As certain plants and animals become increasingly rare, the price people are willing to pay will (Increase, Decrease).

26. Exploiters of natural biota (Encourage, Discourage) public education on their assault on natural biota.

Combination of Factors

27. Indicate which of the below factors did [+] or did not [-] cause the decline of many fish and shellfish populations in the Chesapeake Bay.

[ ] pollution
[ ] habitat destruction
[ ] overuse
[ ] introduction of foreign species

Conclusion and the Tragedy of the Commons

28. Humans appear to value wildlife trinkets, pollution, and extravagant land use (More, Less) than a sustainable biosphere.

29. Which of the following actions or statements are [+] or are not [-] characteristic of the *tragedy of the commons*?

[ ] Whoever grazes the most cattle gains an economic advantage over those who graze less.
[ ] "If I don't harvest this resource, someone else will."
[ ] Reacting to diminished catches of lobsters by fishing less.
[ ] Whenever two or more independent groups are engaged in exploitation of a resource.

CONSERVATION OF NATURAL BIOTA AND ECOSYSTEMS

GENERAL PRINCIPLES

30. Any resource that has the capacity to renew or replenish itself is called ________________ resource.

31. Is it (True, False) that conservation implies complete denial of any use of natural resources?

32. The aim of conservation is to ________________ or ________________ use.

Concept of Maximum Sustained Yield

33. Match the following definitions with [a] maximum sustained yield and [b] carrying capacity.

[ ] The maximum population that an ecosystem can support.
[ ] The maximum use a system can sustain without impairing its ability to renew itself.

34. Maximum sustained yield is obtained with (Optimal, Maximum) population size.

35. Maximum sustained yield is (Reached, Exceeded) when use begins to destroy regenerative capacity.

36. Indicate whether the following population characteristics will increase [+] or decrease [-] sustained yield.

[ ] A population below or within the carrying capacity.
[ ] A population that is approaching or exceeding carrying capacity.

37. Is it (True, False) that carrying capacity and optimal population size are constant?

38. Indicate whether the following procedures will increase [+] or decrease [-] sustained yield.

[ ] Pollution and other forms of habitat alteration.
[ ] Use levels that ignore population characteristics in relation to carrying capacity.
[ ] Using suitable management procedures.
[ ] Thinning populations that are at or exceeding the carrying capacity.

Specific Points

39. Indicate whether the following procedures will sustain [+] or diminish [-] maximum sustained yield.

[ ] Protecting biota from the "tragedy of the commons".
[ ] Invoking regulations but without enforcement.
[ ] Reducing the economic incentives that promote violation of regulations.
[ ] Preserving habitats.
[ ] Protecting habitats from pollution.

WHERE WE STAND WITH REGARD TO PROTECTING BIOTA

Game Animals in the United States

40. Wildlife and game animals are treated as a ____________________.

41. Restrictions on hunting that protect the wildlife commons are

a. requiring a hunting license. (True, False)
b. limiting the type of weapon used in the harvest. (True, False)
c. hunting within specified seasons. (True, False)
d. setting size and/or sex limits on harvested animals. (True, False)
e. limiting the number that can be harvested. (True, False)
f. prohibiting commercial hunting. (True, False)

42. Suburban or recreational development in rural areas is fragmenting the environment into ecological __________________.

43. Indicate whether the following activities increase [+] or decrease [-] the wildlife commons.

[ ] Establishing game preserves.
[ ] Habitat loss through suburban or recreational development in rural areas.
[ ] Road-killed wildlife.
[ ] "Ecological islands".
[ ] Draining wetlands.

Exotic Species and the Endangered Species Act

44. Numerous species of wildlife have been subjected to commercial hunting for their ________________, ________________, ________________, ________________, or to simply sell as ________________.

45. Snowy egrets were heavily exploited in the 1800s for their (a. meat, b. eggs, c. nests, d. feathers) that were used in (a. fancy restaurants, b. ladies hats, c. curiosity shops, d. grocery stores).

46. The two states that were first to pass laws protecting plumed birds were ________________ and ________________.

47. The law that forbids interstate commerce in illegally killed wildlife is called the ________________ Act.

48. The law that protects species from extinction is called the ________________ ________________ Act of ________________.

49. Any species that has been reduced to the point of imminent danger of becoming extinct is defined as ________________.

50. Indicate whether the following provisions are strengths [+] or weaknesses [-] of the Endangered Species Act.

[ ] The need for *official recognition.*
[ ] Stiff fines for the killing, trapping, uprooting, or commerce of endangered species.
[ ] Controls on government development projects in critical habitats.
[ ] Enforcement of the act.

51. Is it (True, False) that zoos or "seed banks" will be able to maintain the total genetic diversity that existed in the total wild populations of endangered species?

Aquatic Species

52. Open oceans were traditionally considered an ________________ commons.

53. As a response to overfishing in the international commons, many nations extended their ________________ limits.

54. In 1977, the United States extended their territorial limits from 3 to 12 miles to ________________ miles offshore.

55. Indicate whether extending territorial limits has [+] or has not [-] promoted conservation of the following organisms.

[ ] whales, [ ] tuna, [ ] fish and shellfish of freshwater estuaries

56. Today, (Pollution, Overharvest) is the greatest threat to many species and habitats.

IS SAVING SPECIES ENOUGH? THE NEED FOR SAVING ECOSYSTEMS

57. Contemporary management procedures should concentrate on the (Species, Ecosystem).

58. If we save the species and lose the ecosystem, we will lose

a. thousands of other species important to agriculture and medicine. (True, False)
b. nature's services. (True, False)
c. the scientific, recreational, and aesthetic values of natural biota. (True, False)
d. commercial value of natural biota. (True, False)

SAVING TROPICAL FORESTS

59. Tropical rain forests provide

a. millions of plant and animal species. (True, False)
b. a system that maintains the climate of the Earth. (True, False)

60. Indicate whether the following conditions promote [+] or prevent [-] destruction of the tropical rain forest.

[ ] Uncontrolled population growth.
[ ] Huge international debts.
[ ] Fast food chains and cheap hamburger.

WHAT YOU CAN DO TO HELP SAVE NATURAL BIOTA AND ECOSYSTEMS

61. List three actions you can take to save natural biota and ecosystems.

a. ______________________________

b. ______________________________

c. ______________________________

KEY VOCABULARY AND CONCEPTS

natural biota	values of natural biota	cultivar
natural services	overuse	introductions
tragedy of the commons	renewable resource	conservation
maximum sustained yield	carrying capacity	hunting regulations
exotic species	endangered species	Endangered Species Act

SELF TEST

Circle the correct answer to each question.

1. All plants, animals, and microbes which constitute ecosystems are referred to as

a. biomes, b. biomass, c. biota, d. biosphere

2. The use of natural biota for all food, clothing, fuel, and building materials by humans occurred

a. in primitive times, b. 10,000 years ago, c. 250 years ago, d. in the 20th century

3. Values of natural biota do not include

a. aesthetic and recreational enjoyment.
b. provision of natural services.
c. provision of mineral resources.
d. providing a bank of genetic resources.

4. Cultivated (cultivar) populations of plants and animals are different from wild populations in which of the following ways?

a. They are highly adaptable to changing environmental conditions.
b. They have numerous traits for resistance to parasites.
c. They have a low degree of genetic diversity in their gene pools.
d. They represent a reservoir of genetic materials called a genetic bank.

5. Natural biota do not play a significant role in the control of

a. erosion, b. climate, c. air quality, d. volcanic eruptions.

6. A recreational value of natural biota would not normally include

a. hunting, b. sport fishing, c. ecological studies, d. camping

7. Which of the following is often the end result when commercial interests become involved with natural biota resource?

a. preservation, b. destruction, c. conservation, d. managed harvests

8. Draining wetlands for development or agriculture is an assault on natural biota known as

a. loss of habitat, b. pollution, c. overuse, d. introduction of foreign species

9. The growing fad for exotic pets and house plants is an assault on natural biota known as

a. loss of habitat, b. pollution, c. overuse, d. introduction of foreign species

10. Which of the following factors did not cause the decline of many fish and shellfish populations in the Chesapeake Bay?

a. loss of habitat, b. pollution, c. overuse, d. introduction of foreign species

11. Whenever two or more independent groups are engaged in exploitation of a resource with no restrictions, this often leads to

a. a lack of understanding of maximum sustained yield.
b. the "tragedy of the commons".
c. conservation of the resource.
d. consumer concern regarding harvests of renewable resources.

12. Regulations that prevent harvests of natural biota without impairing their ability to renew previous population levels are responses to the concept of

a. maximum sustained yield, b. carrying capacity, c. endangered species, d. all of these

13. Which of the following procedures will decrease sustained yields of natural biota?

a. Thinning populations that are at or exceeding the carrying capacity.
b. Invoking regulations but without enforcement.
c. Protecting biota from the "tragedy of the commons".
d. Reducing economic incentives that promote violation of regulations.

14. The law that specifically addresses species extinctions is called the

a. Lacey Act, b. Conservation Reserve Act, c. Endangered Species Act, d. Multiple Use Act

15. If we save the species and lose the ecosystem, we will also lose

a. thousands of other species important to agriculture and medicine.
b. nature's services.
c. commercial value of natural biota.
d. all of the above

ANSWERS TO STUDY QUESTIONS

1. biota; 2. 1.5, 5 to 10; 3. e, a, b, d, c; 4. true; 5. agriculture, medicine, natural services, aesthetics and recreation, commercial; 6. a, b, a, b, b, a, b, a; 7. treatment for leukemia; 8. rosy periwinkle; 9. true; 10. all +; 11. false; 12. a, a, a, a, b, c, b; 13. increase; 14. decreases; 15. destruction; 16. loss of habitat, pollution, overuse, introduction of foreign species; 17. d; 18. a; 19. d; 20. a; 21. b; 22. a; 23. c; 24. c; 25. increase; 26. discourage; 27. +, +, +, -; 28. more; 29. +, +, -, +; 30. renewable; 31. false; 32. manage, regulate; 33. b, a; 34. optimal; 35. exceeded; 36. -, +; 37. false; 38. -, -, +, +; 39. +, -, +, +, +; 40. commons; 41. all true; 42. islands; 43. +, -, -, -, -; 44. furs, skins, plumage, tusks, pets; 45. d, b; 46. Florida, Texas; 47. Lacey; 48. Endangered Species Act, 1966; 49. endangered; 50. -, +, +, -; 51. false; 52. international; 53. territorial; 54. 200; 55. all -; 56. pollution; 57. ecosystem; 58. all true; 59. all true; 60. all +; 61. prevent sale of exotic wildlife or their parts, support conservation organizations, pursue political support.

ANSWERS TO SELF TEST

1. c; 2. a; 3. c; 4. c; 5. d; 6. c; 7. b; 8. a; 9. d; 10. d; 11. b; 12. a; 13. b; 14. c; 15. d

CHAPTER 18

CONVERTING REFUSE TO RESOURCES

When did you last take out the garbage? How often do you perform this task? It is probably a task that you perform frequently, on a regular basis, and with little thought to what happens to it. Few people spend time thinking deep thoughts about garbage! However, this chapter is going to focus your attention on what does happen to your garbage in terms of how it is disposed and, more importantly, what are the ecological impacts of a society that has a "throw-away" mentality. Imagine your personal reaction if the next designated landfill was your front yard. You would probably object vehemently to such a ridiculous notion. But can you appreciate, in turn, that the landfill you presently use is nature's front yard?

Make a mental list of all the newspapers, appliances, chemicals, cans and bottles, packaging materials, clothing, and food that you have discarded over the past week. Multiply your volume of refuse by each household on your block, by each household in your town, and finally by each household in the United States. The total is an immense tonnage of wastes on a one-way trip! Where does it all go? The brutal truth is that we are on the brink of a disaster because we are ever increasing our production of trash and are rapidly running out of places to put it. What would you do with your pile of trash if the garbage truck did not show up? In this chapter, we will examine the dimensions of the refuse crisis and some possible solutions to the crisis.

STUDY QUESTIONS

THE SOLID WASTE CRISIS

BACKGROUND OF SOLID WASTE DISPOSAL

1. The total of all refuse thrown away from homes and commercial establishments is called municipal ____________________ wastes.

2. Two factors that have contributed to a steady increase in the volume of municipal solid wastes are changing ________________________ and increasing use of __________________ materials.

3. The three largest components of municipal solid waste are ____________________, ____________________, and ____________________.

4. The proportions of the components of refuse change in terms of

   a. the generator, b. the neighborhood, c. time of year, d. all of these factors

5. Grass clippings, leaves, and other lawn wastes are called ______________ wastes.

6. Is it (You, Government) who assumes responsibility for collecting and disposing of municipal solid wastes?

7. The two early forms of municipal solid waste disposal were open ____________________ and ____________________.

PROBLEMS OF LANDFILLS

8. Indicate whether the following examples are related to the landfill problems of [a] leachate generation, [b] methane production, or [c] settling.

   [ ] explosions
   [ ] groundwater contamination
   [ ] development of shallow depressions

IMPROVING LANDFILLS: TRYING TO FIX A WRONG ANSWER

9. Indicate whether the following landfill improvements attempt to correct problems of [a] leachate generation, [b] methane production, or [c] both a and b

   [ ] Siting landfills on high ground well above the water table.
   [ ] Contoured floors to drain water into tiles.
   [ ] Covering the floor of the landfill with 12 inches of impervious clay or a plastic liner.
   [ ] A gravel layer that surrounds the entire fill.
   [ ] Shaping the refuse pile into a pyramid.
   [ ] Monitoring wells.

10. In summary, the majority of landfill improvements attempt to correct (Leachate, Methane, Settling) problems.

ESCALATING COSTS OF LANDFILLING

11. Increasing costs of landfilling are related to

   a. new design features. (True, False)
   b. acquiring a site. (True, False)
   c. transportation. (True, False)

12. Acquiring a landfill site far enough away from suburbia automatically increases ____________________ costs.

13. Indicate whether the following events will [+] or will not [-] happen because of difficulties in establishing new landfill locations.

   [ ] Continued use of old landfills with inadequate safeguards.
   [ ] More old landfills closed than can be replaced with new landfills.
   [ ] A landscape covered with pyramids.
   [ ] Reduced refuse production by society.

SOLUTIONS

14. List five alternatives to landfilling some of the components of refuse.

   a. ____________________, b. ____________________, c. ____________________

   d. ____________________, e. ____________________.

THE RECYCLING SOLUTION

Match the below statements with the following list of impediments to recycling refuse.

[a] sorting
[b] lack of standards
[c] reprocessing
[d] separation between government and private enterprise
[e] marketing
[f] vested interests in the *status quo*
[g] hidden costs

15. [ ] Several kinds of plastics or grades of paper may be used in similar or the same products.

16. [ ] Collection and converting refuse into salable materials.

17. [ ] The constituents of refuse must be separated at home.

18. [ ] The profits generated from indefinite manufacture of throw-away containers.

19. [ ] Identification of industrial or consumer markets.

20. [ ] Local governments collect the refuse and private enterprise produces the refuse.

21. [ ] Few people realize how much they are paying for refuse disposal.

Addressing the Problems

22. List three ways communities are overcoming the above impediments to recycling.

a. ______________________________

b. ______________________________

c. ______________________________

23. Match the below types of refuse with the type of product that reprocessing will produce.

Refuse Type	Product
[ ] paper	a. compost
[ ] glass	b. refabrication without ore extraction
[ ] plastic	c. synthetic lumber
[ ] metals	d. substitute for gravel and sand
[ ] food and yard wastes	e. cellulose insulation
[ ] textiles	f. strengthening agent in recycled paper

Promoting Recycling Through Mandate

24. Which of the below examples of laws that might be passed to promote recycling place the burden mainly on the consumer rather than the government?

a. mandatory recycling laws, b. banning the disposal of certain items in landfills,

c. mandating government purchase of recycled materials, d. advance disposal fees

COMPOSTING

25. Composting produces a residue of decomposition called ________________.

26. Is it (True, False) that there are business opportunities in composting?

REFUSE TO ENERGY CONVERSION

27. Is it (True, False) that refuse can be turned into electricity?

28. Burning raw refuse eliminates what three problems of recycling (see answer choices for questions 15 to 21 above).

    a. ______________________, b. ______________________, c. ____________________.

29. The values of incinerated refuse may include

    a. salvage of metals. (True, False)
    b. extending the life of the landfill. (True, False)
    c. fill dirt in construction sites. (True, False)

REDUCING WASTE VOLUME

30. The most efficient form of recycling is ________________ items in their existing capacity.

31. (Returnable, Nonreturnable) beverage containers is a classic example of reusing items.

32. Larger profits are generated from nonreturnable containers through

    a. reduced transportation costs. (True, False)
    b. indefinite bottle production. (True, False)
    c. eliminating the returnable bottle competition. (True, False)
    d. opposing bottle bills. (True, False)

33. Bottle bills encourage the use of (Returnable, Nonreturnable) containers.

34. Bottle bills will (Decrease, Increase) jobs and (Decrease, Increase) litter.

35. Other measures to reduce the amount of material going into trash include

    a. reducing the amount of material in products, b. downsizing, c. increased durability, d. flea markets and yard sales, e. all of these are trash reduction procedures

INTEGRATED WASTE MANAGEMENT

36. Integrated waste management uses (One, Several) recycling methods.

CHANGING PUBLIC ATTITUDES AND LIFESTYLE - WHAT YOU CAN DO

37. Place an [X] by each of the following actions you have personally taken to reduce solid waste.

    [ ] Learned how solid wastes are disposed of in your community.
    [ ] Calculated the hidden costs for refuse disposal.
    [ ] Encouraged your state delegates to support a bottle bill.
    [ ] Recycled paper, cans, or yard refuse.
    [ ] Maintained a policy of purchasing durable goods.
    [ ] Refused to litter.

KEY VOCABULARY AND CONCEPTS

municipal solid wastes	yard wastes	landfill
open burning dumps	leachate	methane production
settling	pyramids	groundwater monitoring wells
sorting	lack of standards	reprocessing
marketing	vested interests	hidden costs
recycling	advanced disposal fee	composting
waste to energy conversion	returnable bottles	bottle bill
yard sales	downsizing	integrated pest management

SELF TEST

Circle the correct answer to each question.

1. Which of the following has not been a cause of ever increasing volumes of municipal solid wastes?

   a. changing lifestyles, b. disposable materials, c. recycling, d. increased gross national product

2. The three largest categories of municipal solid wastes are

   a. paper, cans, and bottles.
   b. paper, food, and glass.
   c. plastics, paper, and metals.
   d. food, paper, and plastics.

3. The proportions of the components of refuse change in terms of

   a. the generator, b. the neighborhood, c. time of year, d. all of these

4. Most municipal solid waste in the United States is presently disposed of

   a. by burning it in a closed incinerator.
   b. in landfills.
   c. by barging it to sea and dumping it.
   d. in open-burning dumps.

5. The potential of groundwater contamination from landfills is primarily due to

   a. leachate, b. methane gas, c. settling, d. explosions

6. Which of the following landfill improvements attempt to correct problems of methane gas build-up?

   a. Siting landfills on high ground well above the water table.
   b. Surrounding the entire fill with a gravel layer.
   c. Shaping the refuse pile into a pyramid.
   d. Contoured floors to drain water into tiles.

7. Increasing costs of landfilling are related to

   a. new design features, b. acquiring new sites, c. transportation, d. all of these

8. Even though establishing new landfill locations is extremely difficult, which of the following events will not occur in spite of this difficulty?

   a. Continued use of old landfills with inadequate safeguards.
   b. More old landfills closed than can be replaced with new landfills.
   c. Reduced refuse production by society.
   d. A landscape covered with pyramids.

9. Collection and converting refuse into salable materials is what kind of impediment to recycling?

   a. sorting, b. lack of standards, c. marketing, d. reprocessing

10. In term of recycling potential, paper can be

    a. repulped and made into paper and paper products.
    b. manufactured into cellulose insulation.
    c. composted to make a nutrient-rich humus.
    d. all of the above

11. Which of the below examples of recycling laws places the burden mainly on the consumer rather than the government?

    a. mandatory recycling
    b. advance disposal fees
    c. bans on the disposal of certain items
    d. mandating government purchase of recycled materials

12. Burning raw refuse does not eliminate which of the following problems of recycling?

    a. sorting, b. hidden costs, c. reprocessing, d. marketing

13. Which of the following is not a potential value derived from incinerated refuse?

    a. metal salvage, b. returnable bottles, c. extended use of existing landfills,
    d. fill dirt for construction sites

14. Which of the following is not a potential value of "bottle bills"?

a. Reduced transportation costs.
b. Indefinite bottle production.
c. Increased jobs.
d. Decreased litter.

15. Which of the following measures reduce the amount of material going into trash?

a. downsizing, b. decreased obsolescence, c. yard sales, d. all of these

---

ANSWERS TO STUDY QUESTIONS

1. solid; 2. lifestyle, disposable; 3. paper, food, glass; 4. d; 5. yard; 6. government; 7. burning, landfills; 8. b, a, c; 9. a, a, a, b, a, a; 10. leachate; 11. all true; 12. transportation; 13. +, +, +, -; 14. recycling, composting, refuse to energy, reducing waste volume, integrated waste management; 15. b; 16. c; 17. a; 18. f; 19. e; 20. d; 21. g; 22. forming partnerships between government and business, sorting, reprocessing and profits; 23. e, d, c, b, a, f; 24. d; 25. humus; 26. true; 27. true; 28. sorting, reprocessing, marketing; 29. all true; 30. reusing; 31. returnable; 32. all true; 33. returnable; 34. increase, decrease; 35. e; 36. several; 37. hopefully all X's

---

ANSWERS TO SELF TEST

1. c; 2. b; 3. d; 4. b; 5. a; 6. b; 7. d; 8. c; 9. c; 10. d; 11. b; 12. b; 13. b; 14. b; 15. d

CHAPTER 19

ENERGY RESOURCES AND THE NATURE OF THE ENERGY PROBLEM

In order to comprehend the importance of Chapter 19 content, develop an inventory of all your direct and indirect reliance on energy just to maintain your quality of life for one day. For example, consider the energy required to maintain a comfortable environment, produce and prepare food, provide transportation, and to produce the diversity of goods and services we all enjoy. Do you know the source of our energy - what we now consider our energy reserves? It is not the sun or any other sustainable supply. We have near total reliance on fossil fuels as our reserve of potential energy. As you know from previous chapters, fossil fuels are nonrenewable and nonrecyclable. In short, we rely almost totally on an energy reserve that is bound to become exhausted, perhaps in our lifetime. To compound the problem, our energy demands increase proportionally to population increase and consumer demand for new and different products and services. We are clearly on a path to nonsustainable development.

What are the expected outcomes of nonsustainable development? Try to imagine how dramatically your quality of life would change if our energy supplies were interrupted for 48 hours. For most, this would mean no electricity, no heating or air conditioning, no gasoline, no hot baths, no lights, no elevator service, no reason to go to work, and a multitude of other dramatic and negative results. Our energy reserves are one-way trips which means once they are used they are gone forever. One can therefore predict that a quality of life that is dependent on nonrenewable energy reserves has a life span that is directly correlated to the amount of the reserve. When the reserve runs out, the quality of life ends. The purpose of this chapter is to provide a clear picture of the nonsustainability of current energy resources.

STUDY QUESTIONS

ENERGY SOURCES AND USES IN THE UNITED STATES

1. Mechanization allowed the production of (More, Less) food and required (More, Less) energy.

2. There is a (Positive, Negative) correlation between gross national product and energy consumption (see Figure 19-2).

BACKGROUND

3. Throughout human history, the major energy source was _______________ labor.

4. The primary limiting factor to machinery designs in the early 1700s was a _______________ source.

5. The _________________ engine launched the Industrial Revolution in the late 1700s.

6. The first major fuel for steam engines was (a. coal, b. wood, c. gasoline, d. electricity).

7. By the 1800s the major fuel for steam engines was (a. coal, b. wood, c. gasoline, d. electricity).

8. By 1920, coal provided _________ percent of all energy used in the United States.

9. Two 1800 technologies that provided an alternative to coal were __________________ drilling and __________________ combustion.

10. Indicate whether the following attributes are [A] advantages, [D] disadvantages of oil-based fuels when compared to coal.

    [ ] portability, [ ] pollution, [ ] energy content, [ ] known reserves

11. Oil-based fuels running internal combustion engines became the dominant source by (a. 1900, b. 1930, c. 1950, d. 1970).

12. Electricity is a (Primary, Secondary) source of energy.

13. The major fuel for generating electrical power is (a. coal, b. wood, c. oil-based fuels).

14. In the 1960s, ________________ power became a means of generating electrical power.

15. Natural gas (methane) burns (More, Less) cleanly than oil and produces (More, Less) pollution.

16. The rapid increase in energy consumption between 1960 and 1980 (see Figure 19-8) was directly related to population shifts to ___________________ living and ________________ by private cars.

17. The "energy crisis" in the 1970s resulted in the use of (Less, More) energy and the search for (More, Fewer) energy alternatives.

THE CURRENT SITUATION

18. The four main energy sources are

    a. ______________________ (43%), b. ______________________ (25%)

    c. ______________________ (22%), d. ______________________ (10%)

    Match the dominant energy source with each of the four major purposes listed below.

    a. oil-based fuels, b. natural gas, c. coal, d. nuclear power and other sources,
    e. more than one of these energy source is used for the stated purpose

19. [ ] transportation

20. [ ] industrial processes

21. [ ] space heating and cooling

22. [ ] generation of electrical power

23. The energy source that supplies over 40 percent of our total energy and virtually 100 percent of our transportation is (a. oil-based fuels, b. natural gas, c. coal, d. nuclear power and other sources).

24. All nuclear power and virtually all coal is used to generate ________________ power.

25. The U.S. energy policy is one of (Sustainability, Nonsustainability).

THE ENERGY DILEMMA (OR CRISIS): DECLINING RESERVES OF CRUDE OIL

FORMATION OF FOSSIL FUELS

26. Coal, crude oil, and natural gas are called ________________ fuels.

27. Fossil fuels are accumulations of organic matter from rapid ________________ activity that occurred at various times in the Earth's history.

28. Indicate whether the following statements are true [+] or false [-] concerning fossil fuels.

    [ ] Supplies are limited.
    [ ] The significant accumulations of organic matter that produced them can happen today.
    [ ] We are using them far faster than they can be produced.

EXPLORATION, RESERVES, AND PRODUCTION

29. Amounts of crude oil that are estimated to exist in the earth are called estimated ________________.

30. Oil fields that have been located and measured are called ________________ reserves.

31. The maximum rate of production from a given oil field is about __________ percent of the (Proven, Remaining) reserves.

    Assume that a field has 100 million barrels of recoverable reserves. Given a maximum production of 10 percent, how much can be withdrawn in

32. Year 1: ________________ barrels leaving ________________ barrels

33. Year 2: ________________ barrels leaving ________________ barrels

34. Is it (True, False) that production from a given field may continue indefinitely?

35. Is it (True, False) that as maximum production goes down, the price per barrel will increase?

36. As proven reserves are decreased by production, they must be increased by ____________ discoveries.

DECLINING U.S. RESERVES AND INCREASING IMPORTATION OF CRUDE OIL

37. The last major discovery of oil in the United States was in 1968 in ________________.

38. The consumption of crude oil in the United States in 1970 was over ________ billion barrels per year.

39. One barrel of oil equals __________ gallons.

40. Indicate whether the following events have increased [I] or decreased [D] in the United States since the 1970s.

[ ] Consumption of fuels derived from oil.
[ ] Discoveries of new oil in the United States.
[ ] Production of oil in the United States.
[ ] The gap between production and consumption.
[ ] United States dependence on foreign oil.

THE CRISIS

41. **OPEC** stands for ____

42. The OPEC nations created an oil "crisis" in the 1970s by (Increasing, Suspending) oil exports.

43. Through its ability to create world oil shortages, OPEC (Increased, Decreased) the price of crude oil.

RESPONSE TO THE OIL CRISIS OF THE 1970s

44. Indicate whether increasing the price of crude oil caused the following events to increase [I] or decrease [D] in the United States.

[ ] The rate of exploratory drilling and discovery of new oil.
[ ] Production from known fields.
[ ] Efforts toward conservation.
[ ] Consumption.
[ ] Efforts toward development of alternative energy sources.
[ ] Dependence on foreign oil.

45. Is it (True, False) that the results of the oil "crisis" were not shortages but an oil glut?

46. As a result of oil surpluses, the price of a barrel of crude oil (Increased, Decreased).

47. As a result of oil surpluses, the price of a gallon of gas at the gas station did not

a. Increase, b. Decrease, c. Stay at the same level as during the "crisis".

VICTIMS OF SUCCESS: SETTING THE STAGE FOR ANOTHER CRISIS

48. Indicate whether the collapse in oil prices has caused the following events to increase [I] or decrease [D] in the United States.

[ ] The rate of exploratory drilling and discovery of new oil.
[ ] Production from known fields.
[ ] Efforts toward conservation.
[ ] Consumption.
[ ] Efforts toward development of alternative energy sources.
[ ] Dependence on foreign oil.

49. Is it (True, False) that OPEC could again attempt to control oil prices?

50. Is it (True, False) that there will be future shortages of oil?

PREPARING FOR FUTURE OIL SHORTAGES

51. The major preparation for future oil shortages is to (Increase, Decrease) oil consumption.

52. List the two primary ways oil consumption can be decreased.

a. ____________________ and b. ____________________

53. Is it (True, False) that another alternative for preparing to meet future oil shortages is to increase exploration and production?

WHAT YOU CAN DO

54. Place an [X] by any of the following actions you have taken to conserve oil.

[ ] Purchased fuel efficient cars.
[ ] Decreased the commuting distance between your home and work place.
[ ] Maximized the heating and cooling efficiency of your home.
[ ] Considered energy alternative such as solar energy.
[ ] Communicated your concerns about future oil shortages to your elected officials.

KEY VOCABULARY AND CONCEPTS

mechanization
coal
fossil fuels
production
oil field
conservation
sustainable
primary energy source
exploration
estimated reserves
the "crisis"
substitutions
internal combustion
secondary energy source
reserves
proven reserves
OPEC

SELF TEST

Circle the correct answer to each question.

1. A country that consumes a high amount of energy per capita is likely to be

a. an underdeveloped nation with a low standard of living.
b. a highly developed nation with a high standard of living.
c. an underdeveloped nation but the standard of living may be high.
d. a highly developed nation with a low standard of living.

2. The first major fuel for steam engines was

a. coal, b. wood, c. gasoline, d. electricity

3. By the 1800s the major fuel for steam engines was

a. coal, b. wood, c. gasoline, d. electricity

4. Which of the following attributes is a disadvantage of oil-based fuels when compared to coal?

   a. portability, b. pollution, c. energy content, d. known reserves

5. The energy source that supplies over 40 percent of our total energy and virtually 100 percent of our transportation is

   a. oil-based fuels, b. natural gas, c. coal, d. nuclear power

6. The major method of generating electricity in the U.S. involves driving turbines with steam produced by the heat from burning

   a. oil-based fuels, b. natural gas, c. coal, d. nuclear power

7. Which of the following statements is false concerning fossil fuels?

   a. Supplies are limited.
   b. The significant accumulations of organic matter that produced them can happen today.
   c. We are using fossil fuels far faster than they can be produced.
   d. Fossil fuels are accumulations of organic matter from rapid photosynthetic activity.

8. Assume that an oil field has 100 million barrels of recoverable reserves. Given a maximum production of 10 percent, how many barrels can be withdrawn the first year?

   a. 9 million, b. 10 million, c. 90 million, d. 100 million

9. The last major discovery of oil in the United States was in 1968 in

   a. Alaska, b. Texas, c. Oklahoma, d. Gulf of Mexico

10. Which of the following events has been steadily decreasing in the United States since the 1970s?

    a. Consumption of fuels derived from oil.
    b. Discoveries of new oil.
    c. The gap between production and consumption.
    d. United States dependence on foreign oil.

11. The "energy crisis" in the U.S. during the 1970s was basically a result of

    a. suspension of oil exports by OPEC nations.
    b. increased prices on crude oil exports by OPEC nations.
    c. a world wide shortage of oil.
    d. oil companies holding back on production.

12. Which of the following events did not increase in the United States as a result of OPEC nations increasing the price of crude oil?

    a. Efforts toward conservation.
    b. Consumption
    c. Dependence on foreign oil
    d. The rate of exploratory drilling and discovery of new oil.

13. Which of the following events decreased in the United States as a result of the collapse of oil prices?

    a. Consumption
    b. Dependence on foreign oil
    c. The price of a gallon of gasoline at the gas pumps
    d. All of the above increased

14. When and if energy becomes limited we can expect limits on

    a. transportation.
    b. jobs due to limited industrial production.
    c. accessing other energy alternatives.
    d. all of the above

15. The long term (100 years plus) energy needs of this country, regardless of environmental consequences, can be met through

    a. coal, b. nuclear power, c. water power, d. oil reserves

---

ANSWERS TO STUDY QUESTIONS

1. more, more; 2. positive; 3. human; 4. power; 5. steam; 6. b; 7. a; 8. 80; 9. oil, internal; 10. A, A, A, D; 11. c; 12. secondary; 13. a; 14. nuclear; 15. more, less; 16. suburban, commuting; 17. less, more; 18. oil-based, natural gas, coal, nuclear and other; 19. a; 20. e; 21. e; 22. e; 23. a; 24. electrical; 25. nonsustainability; 26. fossil; 27. photosynthetic; 28. +, -, +; 29. reserves; 30. proven; 31. 10; 32. 10 million, 90 million; 33. 9 million, 81 million; 34. true; 35. true; 36. new; 37. Alaska; 38. 5; 39. 42; 40. I, D, D, I, I; 41. Organization of Petroleum Exporting Countries; 42. suspending; 43. increased; 44. I, I, I, I, I, D; 45. true; 46. decreased; 47. decrease; 48. D, D, D, I, D, I; 49. true; 50. true; 51. decrease; 52. conservation, substitutions; 53. false; 54. hopefully, you put an X by all actions

---

ANSWERS TO SELF TEST

1. b; 2. b; 3. a; 4. d; 5. a; 6. c; 7. b; 8. b; 9. a; 10. b; 11. a; 12. c; 13. d; 14. d; 15. a

CHAPTER 20

NUCLEAR POWER, COAL, AND SYNTHETIC FUELS

You have probably now come to realize that we cannot rely solely on oil-based fuels to maintain our quality of life into the future. In fact, it would be both foolhardy and disastrous to perpetuate such a narrow energy policy. Now that you have been convinced of the perils facing an oil-based society, it is fair to ask, "What are the alternatives?" It would be wonderful to give society a long shopping list of energy alternatives, but given our present technology, this is not possible. We have several alternative technologies that are viable only because (1) they are linked to energy producing resources that are more abundant than oil and (2) they can only be used to produce electricity. These alternatives are of little value in the transportation industry unless automobiles can be designed to run on different types of energy reserves, (e.g., electricity). There are other limitations to these alternatives.

Nuclear power was, at one time, considered the panacea for all our energy needs. This unguarded optimism was the result of no long-term experimentation on the promise and perils of nuclear power. Today, we are well aware of the extreme danger that can be reaped upon whole populations and ecosystems given a cavalier approach to the safety requirements in the use of nuclear power (e.g., Chernobyl and Three Mile Island). In fact, people now have almost a paranoid fear of nuclear reactors. This is unfortunate because nuclear power does have great potential as an alternative means of generating electricity. This potential will be realized, even accepted, if the safety technology becomes as well developed as the power generating technology.

In terms of proven reserves only, coal is the alternative energy resource that would give the United States a high degree of independence from the pricing games played by OPEC nations. However, coal fired generating plants produce the majority of our electrical power but also the majority of environmental pollutants. As with nuclear power plants, these pollution dangers need to be corrected before coal can take an accepted place in the energy industry. The purpose of this chapter is to investigate the potential of nuclear power and coal.

STUDY QUESTIONS

NUCLEAR POWER: A DREAM OR DELUSION

1. In the 1960s and early 1970s, the perception of nuclear power plants was one of unguarded (Optimism, Pessimism).

2. Since 1975, the perception of nuclear power plants was one of (Optimism, Pessimism).

3. What factor will have the greatest impact on determining future developments of nuclear power plants?

   a. cost, b. government financing, c. public opinion, d. foreign investments

4. Is it (True, False) that the majority of the Department of Energy's budget still goes to nuclear research and development?

5. Is it (True, False) that impending oil shortages are real?

NUCLEAR POWER

How it Works

6. Which of the following is a definition of [a] fission and [b] fusion?

[ ] A large atom of one element is split into two smaller atoms of different elements.
[ ] Two small atoms combine to form a larger atom of a different element.

7. In fission and fusion, the mass of the products is (More, Less) than the mass of the starting material.

8. The lost mass is converted into tremendous amounts of ________________.

9. *Controlled fission* releases this energy gradually as ________________.

The Fuel for Nuclear Power Plants

10. All nuclear power plants use (Fission, Fusion).

11. The raw material of nuclear power plants is uranium (235, 238).

12. Uranium-235 is the isotope of uranium ________.

13. The isotope that will fission is uranium (235, 238).

14. The "bullet" that starts the uranium-235 fission process is

a. a proton, b. a neutron, c. an electron, d. an atom

15. The end products of a uranium-235 fission are ________________ and two or three ________________.

16. A ________________ reaction occurs when the neutrons cause other fissions which release more neutrons which cause other fissions.

17. The majority of all uranium is the (235, 238) isotope.

18. Enrichment is the process of enhancing the concentration of uranium (235, 238).

19. A nuclear bomb explosion is the result of a (Controlled, Uncontrolled) fission of (High, Low) level of uranium 235 enrichment.

The Nuclear Reactor

20. A nuclear reactor is designed to

a. sustain a continuous chain reaction. (True, False)
b. prevent amplification into a nuclear explosion. (True, False)
c. consist primarily of an array of fuel and control rods. (True, False)
d. make some material intensely hot. (True, False)
e. convert heat into steam. (True, False)
f. convert steam into electricity. (True, False)

21. Indicate whether the following characteristics pertain to [F] fuel or [C] control rods.

[ ] Contain the mixture of uranium isotopes.
[ ] Contain a neutron-absorbing material.
[ ] Starts the chain reaction.
[ ] Controls the chain reaction.
[ ] Becomes intensely hot.

The Nuclear Power Plant

22. A nuclear power plant is designed to

a. convert heat into steam. (True, False)
b. use steam to drive turbogenerators. (True, False)
c. convert steam into electricity. (True, False)
d. prevent meltdown. (True, False)
e. produce super heated water in a reactor vessel. (True, False)

23. The fissioning of a pound of uranium fuel releases energy equivalent to burning (1, 10, 1000) tons of coal.

Environmental Advantages of Nuclear Power

24. Indicate whether the following characteristics pertain to [C] coal-fired or [N] nuclear power plants.

[ ] Requires 3.5 million tons of raw fuel.
[ ] Requires 1.5 tons of raw material.
[ ] Will emit over 10 million tons of carbon dioxide into the atmosphere.
[ ] Will emit no carbon dioxide into the atmosphere.
[ ] Will emit over 400 thousand tons of sulfur dioxide into the atmosphere.
[ ] Will emit no acid forming pollutants.
[ ] Will produce about 100 thousand tons of ash.
[ ] Will produce about 2 tons of radioactive wastes.
[ ] Presents the possibility of catastrophic accidents, e.g., meltdown.

RADIOACTIVE MATERIALS AND THEIR HAZARD

25. One fission product is newly formed atoms called ________________ isotopes.

26. Unstable isotopes gain stability by ejecting ________________ particles or high-energy ________________.

27. Subatomic particles and high-energy radiation is referred to as ______________________ emissions.

28. The unstable isotopes are known as ____________________ substances.

29. Is it (True, False) that other materials in and around the reactor may become unstable isotopes?

30. Unstable isotopes are (Direct, Indirect) products of fission.

31. The direct and indirect products of fission are the __________________ wastes of nuclear power.

32. Radioactive emissions may, depending on dosage,

a. damage biological tissue. (True, False)
b. block cell division. (True, False)
c. cause the immediate death of living organisms. (True, False)
d. damage DNA molecules. (True, False)
e. elevate risks of cancer and birth defects. (True, False)

33. Radioactive emissions from nuclear power plants that are operating normally are generally (Higher, Lower) than background radiation.

Disposal of Radioactive Wastes

34. The process whereby unstable isotopes eject particles and radiation is known as ________________ decay.

35. The main variable that determines the rate of radioactive decay of a given isotope is its ________________ life.

36. Half-lives vary from a fraction of a ________________ to ________________ of years.

37. Which of the following isotopes has the shortest [S] and which has the longest [L] half-life? (see Table 20-1)

[ ] Iodine-131, [ ] Cesium-137, [ ] Plutonium-239

38. Which of the above isotopes would require the shortest containment time?
____

39. Which of the above isotopes would require the longest containment time?
____

40. Generally, (1, 10, 100) half-lives are required to reduce radiation levels to insignificant levels.

41. To be safe, plutonium-239 (half-life = 24,000 years) would require _________ thousand years of containment.

42. Radioactive waste disposal falls into the two categories of ____________________ term and ____________________ term containment.

43. The major problems of radioactive waste disposal are related to

a. finding long-term containment sites. (True, False)
b. transport of highly toxic radioactive waste across the United States. (True, False)
c. the lack of any resolution to the radioactive waste problem. (True, False)

The Potential for Accidents

44. Is it (True, False) that nuclear accidents occur?

45. The catastrophic ramifications of the nuclear accident at Chernobyl were intensified due to the lack of a ____________________ building.

46. Is it (True, False) that a nuclear accident has occurred in the United States?

47. It is estimated that nuclear power plants are now __________ times safer than before the Three Mile Island accident.

ECONOMIC PROBLEMS WITH NUCLEAR POWER

48. Utilities are turning away from nuclear power primarily because of

a. adverse public opinion, b. government constraints, c. increased costs, d. concern for the environment.

49. The cost of building a nuclear power plant has escalated due to

a. new safety standards. (True, False)
b. construction delays. (True, False)
c. shorter-than-expected lifetime of nuclear power plants. (True, False)
d. embrittlement. (True, False)
e. the demand for electricity being met by coal-fired power plants. (True, False)

MORE ADVANCED REACTORS

50. Indicate whether the following are characteristics of breeder [BR], fusion [FU], or both [BO] types of reactors.

[ ] Creates more fuel than it consumes.
[ ] The raw material is uranium-238.
[ ] Splits atoms.
[ ] Produces radioactive wastes.
[ ] Produces plutonium-239 as radioactive waste.
[ ] Fuses atoms.
[ ] Releases energy.
[ ] The raw material is deuterium and tritium.
[ ] The source of unprecedented thermal pollution.

COAL AND SYNTHETIC FUELS

51. The coal reserves in the United States could provide its needs for energy for about (10, 100, 1000) years.

52. The most practical method of coal extraction is __________________ mining.

53. Strip-mining when compared to underground mining

a. allows more of the coal to be extracted. (True, False)
b. temporarily destroys the surface environment. (True, False)
c. requires land reclamation. (True, False)

54. Coal-derived fuels are called ____________________ fuels.

55. Production of synfuels promises to be an exceedingly (Clean, Dirty) industry.

56. Another potential source of synfuels is oil ________________.

57. Oil shale contains a tarlike organic material called _________________.

58. A ton of high-grade oil shale will yield little more than (1/2, 1, 2) barrels of oil.

59. Extraction of oil from oil shale is economically (Practical, Impractical).

60. Extraction of oil from oil shale is environmentally (Clean, Dirty).

KEY VOCABULARY AND CONCEPTS

fission	fusion	isotopes
controlled fission	chain reaction	enrichment
nuclear reactor	fuel elements	fuel rods
fuel assembly	control rods	nuclear power plant
reactor vessel	meltdown	containment building
radioactive emissions	radioactive substances	radioactive wastes
background radiation	radioactive decay	half-life
short-term containment	long-term containment	Chernobyl
Three Mile Island	inherently safe reactors	embrittlement
decommission	breeder reactors	uranium-235
uranium-238	plutonium-239	fusion reactor
hydrogen	deuterium	tritium
lithium	synthetic fuels	synfuels
oil shale	kerogen	

SELF TEST

Circle the correct answer to each question.

1. Which of the below factors will have the greatest impact on determining future developments of nuclear power?

   a. cost, b. government financing, c. public opinion, d. foreign investments

2. Which of the following best describes nuclear fission?

   a. A large atom of one element is split into 2 smaller and different atoms with the resulting release of free neutrons and energy.
   b. Two small atoms are melted together to form a larger atoms with some mass being converted to energy.
   c. Energy is used to split a large atom into 2 smaller atoms with resulting release of energy.
   d. All of the above are proper explanations of fission.

3. The fuel for present nuclear power reactors is

   a. uranium-235, b. uranium-238, c. plutonium-239, d. thorium

4. The "bullet" that starts the uranium-235 fission process is

   a. a proton, b. a neutron, c. an electron, d. an atom

5. What is the process called that results in the splitting of one uranium atom that releases free neutrons causing other uranium atoms to split which in turn releases more neutrons causing other uranium atoms to split?

   a. nuclear fusion, b. nuclear fission, c. chain reaction, d. enrichment

6. The process of enhancing the concentration of uranium-235 is called

   a. nuclear fission, b. a meltdown, c. a chain reaction, d. enrichment

7. Which of the following is not a function of the nuclear reactor?

   a. Sustain a continuous chain reaction.
   b. Prevent amplification into a nuclear explosion.
   c. Convert heat into steam.
   d. Make some material intensely hot.

8. Which of the following is not a function of the fuel rod?

   a. Contain a mixture of uranium isotopes.
   b. Start a chain reaction.
   c. Control a chain reaction.
   d. Become intensely hot.

9. A nuclear power plant is designed to

   a. produce super heated water.
   b. convert heat into steam.
   c. convert steam into electricity.
   d. do all of the above.

10. Which of the following is a disadvantage of nuclear power plants when compared to coal-fired power plants?

    a. The amount of raw material required.
    b. The type of waste produced.
    c. The wattage of electrical power produced.
    d. The amount of sulfur dioxide produced.

11. The major disadvantage(s) of using atomic energy to generate electricity is/are that

    a. it causes a large amount of thermal pollution.
    b. it produces very dangerous waste products.
    c. the ultimate goal of putting a nuclear power plant on line is redundant to the goal of coal-fired generating plants.
    d. All of the above are disadvantages.

12. Radioactive emissions are not know to cause

    a. damage to biological tissues.
    b. damage to DNA molecules.
    c. a greenhouse effect.
    d. increases in background radiation.

13. The major problems of radioactive waste disposal is not related to

a. finding long-term containment facilities.
b. transport of highly toxic radioactive wastes across the United States.
c. a lack of resolution to the radioactive waste problem.
d. background radiation around nuclear power plants.

14. Escalating costs of building a nuclear power plant are not related to

a. environmental concerns.
b. new safety standards.
c. construction delays.
d. embrittlement.

15. The coal reserves in the United States could provide its energy needs for about

a. 10 years, b. 50 years, c. 100 years, d. 500 years

ANSWERS TO STUDY QUESTIONS

1. optimism; 2. pessimism; 3. c; 4. true; 5. true; 6. a, b; 7. less; 8. energy; 9. heat; 10. fission; 11. 235; 12. 238; 13. 235; 14. b; 15. heat, neutrons; 16. chain; 17. 238; 18. 235; 19. uncontrolled, high; 20. true, true, true, true, false, false; 21. F, C, F, C, F; 22. all true; 23. 1000; 24. C, N, C, N, C, N, C, N, N; 25. unstable; 26. subatomic, radiation; 27. radioactive; 28. radioactive; 29. true; 30. indirect; 31. radioactive; 32. all true; 33. lower; 34. radioactive; 35. half; 36. seconds, thousands; 37. [S] = iodine-131, [L] = plutonium-239; 38. iodine-131; 39. plutonium-239; 40. 10; 41. 24; 42. short, long; 43. all true; 44. true; 45. containment; 46. true; 47. 6; 48. a; 49. all true; 50. BR, BR, BR, BO, BR, FU, BO, FU, FU; 51. 100; 52. strip; 53. all true; 54. syn; 55. dirty; 56. shale; 57. kerogen; 58. 1/2; 59. impractical; 60. dirty

---

ANSWERS TO SELF TEST

1. c; 2. a; 3. a; 4. b; 5. c; 6. d; 7. c; 8. c; 9. d; 10. b; 11. d; 12. c; 13. d; 14. a; 15. c

CHAPTER 21

SOLAR ENERGY
OTHER "RENEWABLE" ENERGY SOURCES, AND CONSERVATION

Here is the ultimate trivia question. Can you identify an energy source that is abundant, everlasting, and nonpolluting? You could ask this question of someone while standing outside on a bright sunny day and they would not be able to identify the sun as the correct answer. It is indeed ironic that, given the tremendous energy needs of our society both now and in the future, we have failed to utilize the one energy source that meets the very criteria we seek. The main barrier to full-scale utilization of solar energy is economic. No one knows how to make a profit from selling sunlight! Does this surprise you? You may have thought that the economic barrier was related to development of the technologies to trap and store solar energy. Your study of Chapter 21 will make you aware of several technologies that are quite efficient, practical, and sophisticated. It is reasonable to assume that stock holders in the oil, coal, or nuclear industries would become quite concerned if the public demanded more government support for solar technology development.

The economic worries of stock holders in the "dirty" power industries will be intensified as greater attention is given to the "clean" alternatives of wind, geothermal, tidal, and wave power. It is amazing that these natural power sources have always been available but drastically underutilized. The main reason they have been ignored is the same as for solar energy. How do you make a profit from the wind and waves? This chapter examines the range of technologies that are available and what is necessary to promote their development and implementation.

STUDY QUESTIONS

SOLAR ENERGY

1. Solar energy is a (Potential, Kinetic) form of energy originating from (Chemical, Thermonuclear) reactions in the sun.

2. Solar energy is a (Renewable, Nonrenewable) form of energy.

3. The hurdles to overcome in using solar energy include

   a. absorbing it over a broad area. (True, False)
   b. concentrating and converting it to useful forms. (True, False)
   c. storage. (True, False)

4. Production of biomass through photosynthesis is a (Direct, Indirect) use of solar energy.

5. Capture of light energy using a physical device is a (Direct, Indirect) use of solar energy.

USE OF DIRECT SOLAR ENERGY

6. List six direct uses of solar energy.

   a. ____________________, b. ____________________, c. ____________________

   d. ____________________, e. ____________________, f. ____________________

Space and Water Heating

7. The use of gas heat to heat water is an extremely (Efficient, Inefficient) use of energy.

8. An/a (Active, Passive) solar heating system requires pumps and blowers to distribute heat.

9. An/a (Active, Passive) solar heating system relies on natural convection currents to distribute heat.

10. The most economical means of solar heating is an/a (Active, Passive) system.

11. Indicate whether the functions of the following solar heating components are [a] absorption and conversion, [b] transport, or [c] storage of solar energy.

    [ ] black surface of flat-plate collectors
    [ ] brick walls in the house
    [ ] circulation system
    [ ] rock "reservoir"

12. Utilization of solar energy for space and water heating is generally (Supported, Rejected) by utility companies.

13. Utilization of solar energy for space and water heating is extremely cost- (Effective, Ineffective).

Solar Production of Electricity

14. List two ways sunlight can be used to produce electrical power.

    a. ____________________ and b. ____________________

15. A device that converts light directly to electricity is the ____________________ cell.

16. The photovoltaic or solar cell consists of two layers of

    a. water, b. electrons, c. atoms, d. mylar film.

17. The top layer contains atoms with (Additional, Missing) electrons in the outer orbital.

18. The bottom layer contains atoms with (Additional, Missing) electrons in the outer orbital.

19. The top layer will readily (Lose, Accept) an electron.

20. The bottom layer will readily (Lose, Accept) an electron.

21. The kinetic energy of sunlight forces a movement of electrons from the top to the bottom layer creating an electrical imbalance or ____________________.

22. Solar cell technology

a. is becoming more cost-effective with increased use. (True, False)
b. will not wear out. (True, False)
c. is already competitive with nuclear power in costs per watt output. (True, False)
d. can provide power for anything from a watch to a whole city. (True, False)

Power Towers

23. Power towers use mirrors to convert solar energy to ________________ which runs ________________ that create electricity.

Solar Ponds

24. Solar ponds consist of a layer of ________________ water overlaid with ________________ water.

25. Match the following functions of the solar pond [a] brine water and [b] fresh water.

[ ] absorption of sunlight energy.
[ ] conversion of sunlight energy into heat.
[ ] prevents the heat from escaping.

26. The solar pond produces a type of ________________ effect with fluids.

27. The hot brine solution can be

a. used directly to heat buildings. (True, False)
b. converted to electrical power. (True, False)

28. Indicate whether the following are benefits [+] or drawbacks [-] of solar pond technology.

[ ] Area required for collection of sunlight energy.
[ ] Distance of collection site from place where power is used.
[ ] The amount of electrical power generated.

Solar Production of Hydrogen

29. Indicate whether the following are benefits [+] or drawbacks [-] of hydrogen power.

[ ] Can be used in place of natural gas.
[ ] Could replace gasoline in cars.
[ ] Has no waste products except water.
[ ] Requires the production of free hydrogen through electrolysis.

USE OF INDIRECT SOLAR ENERGY

30. List three indirect forms of solar energy.

a. ________________, b. ________________, c. ________________

31. Biomass energy refers to the (Direct, Indirect) use of biomass as fuel.

32. Bioconversion refers to the (Direct, Indirect) use of biomass as fuel.

33. Indicate whether the following are benefits [+] or drawbacks [-] of biomass energy.

[ ] Availability of the biomass resource.
[ ] Access to the biomass resource.
[ ] Public acceptance and utilization of biomass energy.
[ ] Past history of human harvests within a maximum sustained yield.

34. Methane gas is (Direct, Indirect) use of biomass as fuel.

35. Methane gas is generated from cow ________________.

36. Is it (True, False) that one can get more electrical power from cows than from nuclear power?

37. Alcohol production is a (Direct, Indirect) use of biomass as fuel.

38. Alcohol production utilizes (Fermentation, Distillation, Both).

39. A mixture of alcohol and gasoline is called ________________.

40. Indicate whether the following are benefits [+] or drawbacks [-] of gasohol.

[ ] Social consequences
[ ] Food vs fuel
[ ] Environmental consequences
[ ] Cost-effectiveness

Hydropower

41. Hydropower is a (Direct, Indirect) use of solar energy.

42. The earliest form of hydropower was the use of (Natural, Artificial) water falls.

43. The contemporary form of hydropower is the use of (Natural, Artificial) water falls.

44. Artificial water falls are created by building huge ________________.

45. The artificial water falls of huge dams generate substantial ________________ power.

46. Indicate whether the following are benefits [+] or drawbacks [-] of hydroelectric power.

[ ] Level of pollution generated.
[ ] Level of environmental degradation.
[ ] Amount of total energy produced.
[ ] Geographical distribution of energy produced.

Wind Power

47. Wind power is a (Direct, Indirect) use of solar energy.

48. The earliest form of harnessing wind power were wind (Mills, Turbines).

49. The contemporary form of harnessing wind power is wind (Mills, Turbines).

50. Indicate whether the following are benefits [+] or drawbacks [-] of wind power.

[ ] The size limitations on wind turbines.
[ ] The number of megawatts of electricity produced.
[ ] The level of pollution generated.
[ ] The level of environmental degradation.
[ ] Geographical distribution of energy produced.
[ ] Cost-effectiveness

GEOTHERMAL, TIDAL, AND WAVE POWER

GEOTHERMAL ENERGY

51. The heat of the Earth's interior is referred to as ____________________ energy.

52. Indicate whether the following are benefits [+] or drawbacks [-] of geothermal energy.

[ ] If accessible, it is an everlasting energy resource.
[ ] The number of megawatts of electricity produced.
[ ] The level of pollution generated.
[ ] The level of environmental degradation.
[ ] Geographical distribution of energy produced.
[ ] Cost-effectiveness
[ ] Technology required for extraction.

TIDAL POWER

53. Tidal power converts the energy of (Incoming, Outgoing, Both) tides.

54. Indicate whether the following are benefits [+] or drawbacks [-] of tidal power.

[ ] If accessible, it is an everlasting energy resource.
[ ] The number of megawatts of electricity produced.
[ ] The level of pollution generated.
[ ] The level of environmental degradation.
[ ] Geographical distribution of energy produced.
[ ] Cost-effectiveness
[ ] Technology required for energy conversion.

WAVE POWER

55. Indicate whether the following are benefits [+] or drawbacks [-] of wave power.

[ ] If accessible, it is an everlasting energy resource.
[ ] The number of megawatts of electricity produced.
[ ] Geographical distribution of energy produced.
[ ] Cost-effectiveness
[ ] Technology required for energy conversion.

CONCLUSIONS

56. The three alternative energy options that are nearly developed to the stage of large-scale commercialization are

a. ____________________, b. ____________________, and c. ____________________ .

57. The greatest difficulty to overcome in electric car technology is (Energy Storage, Speed).

58. The greatest difficulty to overcome in passive solar heating is (Technology, Existing Buildings).

NEED FOR CONSERVATION

59. There are (200, 300, 400) million cars world wide and __________ percent of these are in the United States.

60. The most acceptable conservation strategy regarding cars is to build (Fewer, More Efficient) cars.

ENERGY POLICY

61. Provisions of the Public Utilities Regulatory and Policies Act of 1978 (PURPA) include

a. purchase of energy by utility companies from independent producers. (True, False)
b. tax breaks on alternative energy investments of individuals and industry. (True, False)
d. federal money for alternative energy research, development, and demonstration projects. (True, False)

62. Is it (True, False) that one can still get a tax deduction for investing in an alternative energy technology?

WHAT YOU CAN DO

63. We should encourage our congress persons to develop energy policies that (Spend More Money, Shift Priorities).

KEY VOCABULARY AND CONCEPTS

solar energy
indirect solar energy
passive solar heating
power towers
electrolysis
bioconversion
fermentation
hydropower
windmills
geothermal energy
electric cars
tax deductions
renewable energy
flat-plate collector
photovoltaic cell
solar ponds
biomass
methane
distillation
hydroelectric power
wind turbines
tidal power
more miles per gallon
direct solar energy
active solar heating
solar cell
hydrogen production
biomass energy
biogas
gasohol
hydroelectric dams
wind farms
wave power
PURPA

SELF TEST

Circle the correct answer to each question.

1. Which of the following is not a renewable energy resource?

   a. sunlight, b. coal, c. wind, d. falling water

2. Which of the following is not a direct use of solar energy?

   a. photovoltaic cells, b. biomass conversion, c. power towers, d. space heating

3. The biggest problem connected with using solar energy is

   a. there is not enough energy inherent in sunlight to be worth its use.
   b. there is no good means of collecting sunlight energy.
   c. the means of storing it are bulky and expensive.
   d. there are no good ways of converting solar energy into profits.

4. The most cost-effective method of using solar energy for home use is through

   a. passive solar systems.
   b. active solar systems.
   c. photovoltaic solar systems.
   d. selling your excess power back to the utility companies.

5. Which of the following is not a present means of using solar energy?

   a. Direct conversion to electrical power.
   b. Heating homes and other buildings.
   c. Production of flammable hydrogen gas.
   d. Production of biomass for conversion to flammable liquids.

6. Solar cell technology

   a. cannot become cost effective even with increased use.
   b. will wear out.
   c. is competitive with nuclear power in costs per watt output.
   d. has severe limitations in power output.

7. Which of the following solar technologies cannot be used directly to produce electricity?

   a. solar cells, b. power towers, c. solar ponds, d. wind turbines

8. The solar technology that utilizes two layers of electrons to produce an electrical current is

   a. solar cells, b. power towers, c. solar ponds, d. flat-plate collectors

9. A major drawback of solar pond technology is

   a. the amount of electrical power generated.
   b. the area required for collection of sunlight energy.
   c. the greenhouse effect.
   d. the conversion of sunlight energy into heat.

10. A major drawback of solar production of hydrogen is

a. its ability to replace our reliance on natural gas.
b. its ability to replace gasoline in cars.
c. its waste products.
d. the ability to produce free hydrogen atoms.

11. Which of the following is not an indirect source of solar energy?

a. wind, b. solar cells, c. hydropower, d. biomass production

12. Hydroelectric power cannot provide a large percentage of power in the future because of the

a. level of thermal pollution generated.
b. amount of energy produced.
c. limited geographic distribution of energy produced.
d. level of air and water pollution generated.

13. A major drawback of wind power is

a. amount of energy produced.
b. size limitations of wind turbines.
c. level of environmental degradation.
d. cost-effectiveness.

14. The major drawback that geothermal energy, tidal and wave power all have in common is

a. the number of megawatts of electricity that could be produced is low.
b. the technology required for energy conversion is not cost-effective.
c. the level of environmental degradation.
d. limitations on long-term dependency as an energy source.

15. In view of the nature of U.S. energy problems and associated environmental problems, which of the following areas should be given a higher priority in research and development funding?

a. fusion power, b. solar production of hydrogen, c. breeder reactors, d. utilization of coal

---

ANSWERS TO STUDY QUESTIONS

1. kinetic, thermonuclear; 2. renewable; 3. all true; 4. indirect; 5. direct; 6. space and water heating, solar electricity, solar cells, power towers, solar ponds, solar production of hydrogen; 7. inefficient; 8. active; 9. passive; 10. passive; 11. a, c, b, c; 12. rejected; 13. effective; 14. solar cells, steam; 15. solar; 16. b; 17. additional; 18. missing; 19. lose; 20. accept; 21. current; 22. all true; 23. steam, turbogenerators; 24. brine, fresh; 25. a, a, b; 26. greenhouse; 27. all true; 28. -, -, +; 29. +, +, +, -; 30. wind, water, biomass; 31. direct; 32. indirect; 33. +, -, +, -; 34. indirect; 35. manure; 36. true; 37. indirect; 38. both; 39. gasohol; 40. all -; 41. indirect; 42. natural; 43. artificial; 44. dams; 45. electrical; 46. +, -, +, -; 47. indirect; 48. mills; 49. turbines; 50. -, +, +, +, +, +; 51. geothermal; 52. +, +, -, -, -, -, -; 53. both; 54. +, +, +, -, -, -, -; 55. +, -, -, -, -; 56. wind, photovoltaic cell, biogas; 57. storage; 58. existing; 59. 27; 60. more efficient; 61. all true; 62. false; 63. shift priorities

---

ANSWERS TO SELF TEST

1. b; 2. b; 3. d; 4. a; 5. c; 6. c; 7. d; 8. a; 9. b; 10. d; 11. b; 12. c; 13. b; 14. b; 15. b

CHAPTER 22

LIFESTYLE, LAND USE AND ENVIRONMENTAL IMPACT

Well, we have finally arrived. The last chapter pulls everything together. All the environmental problems addressed in previous chapters are the direct result of too many people living a lifestyle that promotes ecosystem instability. Lifestyle means "the way we live". Most organisms are satisfied with a lifestyle that includes food, water, shelter, and adequate space. But people are different from all other organisms. Their lifestyles must also include the possession of **things** or the economic power to purchase things at will. What are some of these things we associate with a quality lifestyle? Examples include such things as three cars per family, twenty pairs of shoes, appliances, recreational equipment, summer homes, shopping malls, and a vast assemblage of trinkets that defy any reasonable explanation of purpose. One can only imagine the tremendous amounts of raw materials and energy needed to sustain people's voracious hunger for things.

Our lifestyle specifications concerning shelter and adequate space have led to urban sprawl and the consequent environmental degradation. The irony of urban sprawl is that it recreated the very conditions people were attempting to escape. For example, air pollution levels in rural areas are, in many locations, far worse than in the city. The social consequences of urban sprawl can now be documented and have been found to be as insidious as the environmental impacts. Some communities are experiencing a reverse migration where people are returning to the city and reclaiming the traditions, landmarks, and culture that they left. Understanding this phenomenon and other ways people can and are moderating their lifestyles so that they have less environmental impact is the subject of this chapter.

STUDY QUESTIONS

URBAN SPRAWL: ITS ORIGINS AND CONSEQUENCES

1. Uncontrolled and unplanned migration of people into rural areas is referred to as ________________ sprawl.

2. Urban sprawl is directly associated with increased ownership of (Land, Cars).

3. Urban sprawl began (Before, After) World War II.

THE ORIGIN OF URBAN SPRAWL

4. Indicate whether the following factors enhanced [+], or had no effect [-] on urban sprawl.

   [ ] Ability to purchase cars.
   [ ] Commuting time.
   [ ] New highway development.
   [ ] Commuting distance.
   [ ] Population congestion.
   [ ] The lack of planned development.
   [ ] Low rural tax base.
   [ ] The environmental or social consequences.

ENVIRONMENTAL CONSEQUENCES OF URBAN SPRAWL

Indicate whether the following consequences of urban sprawl are related to

[a] Depletion of Energy Resources
[b] Air Pollution, Acid Rain, and the Carbon Dioxide Greenhouse Effect
[c] Degradation of Water Resources and Water Pollution
[d] Loss of Recreational and Scenic Areas and Natural Services
[e] Loss of Agricultural Land

5. [ ] The sacrifice of about 2.5 million acres per year of the best agricultural land.

6. [ ] Placement of highways through parks or along stream valleys.

7. [ ] Increased runoff and decreasing infiltration over massive areas.

8. [ ] Increased numbers and densities of cars with resultant exhaust.

9. [ ] Increased car production and commuting miles.

SOCIAL CONSEQUENCES OF URBAN SPRAWL

10. The movement of people from the central city into outlying suburbs is called ________________ migration.

11. Exurban migration

a. has profound economic and social impacts. (True, False)
b. segregates the population along economic and/or ethnic lines. (True, False)

Exurban Migration and Economic and Ethnic Segregation

12. Indicate whether the following characteristics do [+] or do not [-] describe the people who are left behind after exurban migration.

[ ] They are the poor, elderly, handicapped, and minority groups.
[ ] They can obtain mortgage loans.
[ ] They reflect a gentrification of society.
[ ] They are comprised of predominantly economically depressed minorities and whites.

Declining Tax Base and Deterioration of Public Services

13. Local governments

a. are responsible for providing societal services. (True, False)
b. operate from budgets derived from property taxes. (True, False)
c. are separate entities in the central city and suburbs. (True, False)

14. Property values are generally (Higher, Lower) in the central city when compared to suburbs.

15. Property taxes are generally (Higher, Lower) in the central city when compared to suburbs.

16. An eroding tax base is characteristic of the (Suburbs, Central City).

17. The quality and quantity of government services is markedly (Better, Worse) in the central city when compared to the suburbs.

18. The infrastructure of the central city is (Deteriorating, Improving).

Segmentation and Desocialization Caused by New Highways

19. The building of new highways

a. benefits both central city and suburb residents. (True, False)
b. brings people closer together. (True, False)
c. promotes the establishment of parks and other natural areas. (True, False)

Loss of Businesses and Employment

20. Indicate whether the following events are likely to increase [+] or decrease [-] as a result of exurban migration.

[ ] Industrial development in the central city.
[ ] Job opportunities in the central city.
[ ] The tax base of the central city.
[ ] The unemployment rate of central city residents.
[ ] Social unrest and crime rates within the central city.
[ ] The level of education offered to and achieved by central city residents.
[ ] Accessibility of jobs outside the central city by its residents.

Vicious Cycle of Urban Sprawl and Inner City Decay

21. As exurban migration increases, the quality of life in the central city will (Increase, Decrease) when compared to the quality of life in the suburbs.

22. Is it (True, False) that people will never return to the central city from the suburbs?

23. Is it (True, False) that some programs are being implemented to revitalize the urban core?

CONCLUSION

24. Suburban sprawl is a (Sustainable, Nonsustainable) system.

MAKING CITIES SUSTAINABLE

25. Is it (True, False) that we can abandon cities and start all over?

CITIES CAN BE BEAUTIFUL

26. Indicate whether the following are [+] or are not [-] descriptions of most American cities.

[ ] Expressions of the spirit of the community and homes for the comfort of the individual.
[ ] Fountains and flower beds and tree-shaded streets and boulevards.
[ ] Countless cafes where a person with little money may spend hours over a cup of coffee and a newspaper.
[ ] The city is the daily chat with the butcher, the baker, the newsman, and numerous other merchants with whom one shops.

27. American cities are generally places where people desire to (Leave, Stay).

28. Is it (True, False) that cities can be beautiful and sustainable?

CITIES CAN BE SUSTAINABLE

29. Indicate whether the city [C] or the [S] suburb is more energy and resource efficient in the below categories.

[ ] Proximity of people to residences, shops, and workplaces.
[ ] Need for cars.
[ ] Heating and cooling of buildings.
[ ] Ability to utilize alternative energy resources.
[ ] Provision and maintenance of water, sewer, and other utilities.
[ ] Provision of resource-related jobs.

SLOWING URBAN SPRAWL AND AIDING URBAN REDEVELOPMENT

30. Is it (True, False) that development must be focused on the city rather than suburbs?

Slowing Urban Sprawl

Indicate whether the following methods of slowing urban sprawl are examples of

[a] Clustered versus Detached Housing
[b] Zoning Laws
[c] Agricultural and Conservation Districts
[d] Growth Management Initiatives
[e] Land Trusts

31. [ ] Citizens approving bond issues to buy and preserve open spaces.

32. [ ] One of the roles of the Nature Conservancy.

33. [ ] Laws that specify and limit the kind of development that can occur in given areas.

34. [ ] Laws that focus and cluster development for more efficient use of land and other resources.

35. [ ] Housing the same number of people while preserving major portions of the natural landscape.

Stemming Urban Blight and Aiding Urban Redevelopment

36. List three methods of stemming urban blight and aiding urban redevelopment.

a. ______________________________

b. ______________________________

c. ______________________________

37. Home improvements usually lead to (Higher, Lower) property taxes.

38. Higher property taxes usually lead to (More, Less) home improvements.

39. Taxes are usually based on the value of the (Buildings, Land).

40. Taxes should be based on the value of the (Buildings, Land) only.

41. Converting buildings to cooperatives results in (More, Less) care and maintenance of buildings.

42. Urban homesteaders

a. can purchase properties at low prices. (True, False)
b. can obtain low-interest, long-term loans. (True, False)
c. convert slum dwellings into commercially valuable property. (True, False)

The Integrated Urban House

43. The integrated urban house

a. is in the wilderness. (True, False)
b. is an urban house that is made environmentally sound and nearly self-sufficient. (True, False)

WHAT YOU CAN DO

44. Place an [X] by any of the following actions you have personally taken to control urban sprawl and the resultant environmental impacts.

[ ] Investigated landuse planning and policies in your region.
[ ] Became involved in zoning issues in your neighborhood.
[ ] Alerted the Nature Conservancy on particular areas that need to be conserved.
[ ] Attempted to protect natural areas in your community.
[ ] Encouraged city and county governments to change the tax code.
[ ] Considered the impact of your personal lifestyle on the issues addressed in this chapter.

KEY VOCABULARY AND CONCEPTS

lifestyle
commuting
gentrification
city infrastructure
clustered development
growth management initiatives
homesteaders
urban sprawl
exurban migration
property taxes
industrial parks
zoning laws
land trusts
integrated urban house
Highway Trust Fund
exurbs
eroding tax base
sustainable
agriculture and conservation districts
Nature conservancy

SELF TEST

Circle the correct answer to each question.

1. The prime factor that initiated suburban growth at the end of World War II and which has supported urban sprawl is

   a. farmers selling land to developers.
   b. widespread ownership of private cars.
   c. decline of the central city.
   d. developers buying farms.

2. Which of the following factors had no effect on the rate of urban sprawl?

   a. ability to purchase cars, b. commuting time, c. commuting distance, d. rural tax base

3. An environmental consequence of urban sprawl was

   a. depletion of energy resources.
   b. air pollution, acid rain, and the greenhouse effect.
   c. loss of agricultural land.
   d. all of the above

4. Which of the following is not a characteristic of the people left behind after exurban migration?

   a. They are the poor, elderly, handicapped, and minority groups.
   b. They have high levels of education and income.
   c. They are predominantly economically depressed minorities and whites.
   d. They reflect the gentrification of society.

5. The factor that led most directly to a decline of the central cities was because

   a. the most affluent people moved out.
   b. the rural poor moved into the cities.
   c. tax revenues of the cities declined.
   d. the welfare costs of the city increased.

6. Building new highways

   a. benefits both central city and suburb residents.
   b. brings people closer together.
   c. promotes the establishment of parks and other natural areas.
   d. tends to disrupt existing communities.

7. Which of the following events is likely to increase as a result of exurban migration?

   a. Industrial development in the central city.
   b. Job opportunities in the central city.
   c. Social unrest and crime rates within the central city.
   d. Accessibility of jobs outside the central city to its residents.

8. City governments obtain the largest portion of their operational money from

   a. sales tax, b. income tax, c. property tax, d. equal amounts from each tax

9. Which of the following is likely to increase as a result of exurban migration?

a. The level of education offered to and achieved by central city residents.
b. The types and efficiency of local government services.
c. The tax base of the central city.
d. The unemployment rate of central city residents.

10. In which of the following ways is the suburb more energy and resource efficient than the central city?

a. Proximity of people to residence, shops, and workplaces.
b. Heating and cooling of buildings.
c. Provision and maintenance of water, sewer, and other utilities.
d. None of the above.

11. Future development is most likely to emphasize rebuilding the cities because

a. limits of land and energy will preclude further exurban development.
b. society cannot tolerate further decline of the cities.
c. cities are inherently more resource and energy efficient than are suburbs.
d. all of the above

12. Laws that specify and limit the kind of development that can occur in given areas is a method of slowing urban sprawl called

a. zoning laws, b. land trusts, c. growth management initiatives, d. agricultural districts

13. The role of the Nature Conservancy that decreases the rate of urban sprawl is to establish

a. zoning laws, b. land trusts, c. growth management initiatives, d. agricultural districts

14. The most effective way to encourage the central city resident to stem urban blight is to

a. provide mechanisms that give them ownership in their home and community.
b. give them large sums of money to purchase goods and services.
c. have volunteers come in and fix things up for them.
d. use government funds to build large multiple housing units.

15. Which of the following is a method of giving central city residents ownership in their home and community?

a. Changing taxing procedures to include only the land and not the buildings on it.
b. Initiate tenant cooperatives.
c. Promote urban homesteading.
d. all of the above

ANSWERS TO STUDY QUESTIONS

1. urban; 2. cars; 3. after; 4. +, +, +, -, +, +, +, -; 5. e; 6. d; 7. c; 8. b; 9. a; 10. exurban; 11. all true; 12. all +; 13. all true; 14. lower; 15. higher; 16. central city; 17. worse; 18. deteriorating; 19. all false; 20. -, -, -, +, +, -, -; 21. decrease; 22. false; 23. true; 24. nonsustainable; 25. false; 26. all -; 27. leave; 28. true; 29. all C; 30. true; 31. d; 32. e; 33. b; 34. c; 35. a; 36. tax the property not the buildings, form tenant cooperatives, urban homesteading; 37. higher; 38. less; 39. buildings; 40. land; 41. more; 42. all true; 43. false, true; 44. hopefully, you placed an X by all the listed actions

---

ANSWERS TO SELF TEST

1. b; 2. c; 3. d; 4. b; 5. c; 6. d; 7. c; 8. c; 9. d; 10. d; 11. d; 12. a; 13. b; 14. a; 15. d